Dieter Kind

An Introduction to

High-Voltage Experimental Technique

Textbook for Electrical Engineers

With 181 Fig.

Vieweg

CIP-Kurztitelaufnahme der Deutschen Bibliothek

Kind, Dieter:
An introduction to high-voltage experimental technique: textbook for electr. engineers. – 1. ed. – Braunschweig: Vieweg, 1978.
Einheitssacht.: Einführung in die Hochspannungs-Versuchstechnik ⟨engl.⟩
ISBN 978-3-528-08383-0 ISBN 978-3-322-91763-8 (eBook)
DOI 10.1007/978-3-322-91763-8

1978

Satz: Vieweg, Braunschweig

ISBN 978-3-528-08383-0

Preface

High-voltage technology is a field of electrical engineering the scientific principles of which are essentially found in Physics and which, by its application, is intimately linked with industrial practice. It is concerned with the physical phenomena and technical problems associated with high voltages.

The properties of gases and plasmas, as well as liquid and solid insulating materials, are of fundamental significance to high-voltage technology. However, despite all progress, the physical phenomena observed in these media can only be incompletely explained by theoretical treatment, and so experiment constitutes the foreground of scientific research in this field. Teaching and research in high-voltage technology thus rely mainly upon experimental techniques when dealing with problems.

Recognition of this fact is the conceptual basis for the present book. It is primarily intended for students of electrical engineering and aims to provide the reader with the most important tools for the experimental approach to problems in high-voltage technology. An attempt has been made here to indicate important practical problems of testing stations and laboratories, and to suggest solutions. The book should therefore also prove to be a help to the work of the practising engineer.

The theoretical considerations are correlated with the experiments of a high-voltage practical course, which are described in great detail. The treatment assumes as much familiarity with the subject as may be expected from 3rd year students of electrical engineering.

The development of high-voltage technology reaches as far back as the early years of the 20th century. Meanwhile, numerous new branches of electrical engineering exist of which every electrical engineer must possess some knowledge. This development necessarily led to reconsideration of the common scientific principles of electrical engineering, and this naturally influenced the traditionally based terminology of high-voltage technology. Physical quantities have been given throughout in the international system of units "SI".

As far as 30 years back, my esteemed predecessor Prof. Dr.-Ing. E.h. *Erwin Marx* treated the subject matter of this book in his "Hochspannungspraktikum", a book widely circulated in Germany and abroad. In those days the highest transmission voltage was 220 kV, today overstepping the 1 MV mark is within reach. The fact that in the meantime the development of high-voltage technology has continued in leaps and bounds justifies thorough revision of the same material.

This is an updated English version of the German book "Einführung in die Hochspannungs-Versuchstechnik", which was published in its first edition in 1972 as a result of long years of experience, both in teaching and research, at the Technische Universität Braunschweig. Numerous colleagues in the High-Voltage Institute of this university made substantial contributions to the contents of the book as well as to the planning and verification of the described experiments. Particular thanks go to Dr. *Walter Steudle* for his revision of the manuscript and to Mr. *Hans-Joachim Müller* for his exemplary preparation of the drawings.

As far as the present English version is concerned, I wish to express my sincere gratitude to my colleague Dr.-Ing. *Narayana Rao*, Indian Institute of Technology, Madras, who carried out the translation as an experienced scientist and engineer. His work was supplemented by Mrs. *C. C. J. Schneider* M. A. (Cantab), who carefully revised the whole manuscript. Thanks are also due to Dr. *Tim Teich*, UMIST, Manchester, for his competent help, and to the publishers Vieweg-Verlag for their understanding readiness to comply with special requests.

The 1972 edition of the book has meanwhile been well received. It is my sincere wish that this English edition may now also become a modest contribution to the progress in high-voltage technology beyond the German speaking countries.

Dieter Kind

Contents

List of Symbols Used

a	Length
b	Width, atmospheric pressure, mobility
c	Velocity of light, length
d	Diameter, relative air density
f	Frequency
i	Current (instantaneous value), running index
k	Proportionality factor
l	Length
m	Mass, natural number
n	Natural number, pulse rate, charge carrier density
p	Pressure, Laplace operator
q	Charge
r	Radius, spacing
s	Gap distance, standard deviation
s(t)	Step function
t	Time
u	Voltage (instantaneous value)
ü	Open circuit transformation ratio
v	Velocity, coefficient of variation
w(t)	Unit step response
x	Local coordinate
y	Local coordinate
A	Area, constant
B	Magnetic induction, constant
C	Capacitance
C'	Capacitance per unit length
D	Dielectric displacement, diameter
E	Electric field strength
F	Force, formative area, switching gap
G(p)	Transfer function
I	Current (fixed value)
$\bar{I}$	Current (arithmetic mean value)
$\hat{I}$	Current (peak value)
K	Constant
L	Self inductance
L'	Inductance per unit length

M Mutual inductance
N Number of turns
P Power, probability, test object
P' Power density
Q Electric charge, heat quantity
R Resistance, radius
S Current density, rate of voltage rise (steepness)
T Periodic time, time constant, response time
U Voltage (fixed value)
U_{rms} Voltage (r.m.s. value)
$\bar{U}$ Voltage (arithmetic mean value)
$\hat{U}$ Voltage (peak value)
W Energy
W' Energy density
X Reactance
Y Admittance
Z Surge impedance, apparent resistance
α Ionisation coefficient, abbreviation
β Angle
δ Loss angle
$\tan\delta$ Dissipation factor
ϵ Dielectric constant, earthing coefficient
η Utilization factor
ϑ Temperature
κ Conductivity
μ Permeability
ν Running index
ρ Specific resistance
σ Surface charge density
τ Transit time
φ Electric potential
ω Angular frequency
Φ Magnetic flux

Miscellaneous Symbols

CM Measuring capacitor
KF Sphere gap
KO Cathode-ray oscilloscope

KS Short-circuiting switch
IEC International Electrotechnical Commission[1])
PD Partial discharge
SK Commutating switch
SM Peak voltage measuring device
T Testing transformer
V Diode

---

[1]) The most relevant recommendations here of this body are the following:
IEC TC 42
IEC Publication 60, Parts 1 and 2 (1973)
Part 3 (1976)
Part 4 (1977)

1 Fundamental Principles of High-Voltage Experimental Techniques

1.1 Generation and Measurement of High Alternating Voltages

High alternating voltages are required in laboratories for experiments and a.c. tests as well as for most of the circuits for the generation of high direct and impulse voltages. Test transformers generally used for this purpose have considerably lower power rating and frequently much larger transformer ratios than power transformers. The primary current is usually supplied by regulating transformers fed from the mains supply or, in special cases, by synchronous generators.

Most tests and experiments with high alternating voltages require precise knowledge of the value of the voltage. This demand can normally only be fulfilled by measurements on the high-voltage side of the supply; various techniques for the measurement of high alternating voltages have been devised for this purpose.

1.1.1 Characteristic Parameters of High Alternating Voltages

The shape of u(t) for high alternating voltages will often deviate considerably from the sinusoidal. In high-voltage engineering, the peak value $\hat{U}$ and the root-mean-square (r.m.s.) value

$$U_{rms} = \sqrt{\frac{1}{T}\int_0^T u^2(t)\,dt}$$

are of particular importance.

For high-voltage tests the quantity $\hat{U}/\sqrt{2}$ is defined as the test voltage (IEC-Publ. 60-2, 1973). Here it is assumed that the deviation of the high-voltage shape from the sinusoidal does not exceed the permissible value. For a pure sinusoidal $\hat{U}/\sqrt{2} = U_{rms}$.

Generation of High Alternating Voltages

1.1.2 Test Transformer Circuits

Transformers for generating high alternating test voltages usually have one end of the high-voltage winding earthed. For numerous circuits for the generation of high d.c. and impulse voltages, however, transformers with completely isolated windings are required. Fig. 1.1-1 shows two basic circuits for test transformers. The length of the voltage arrows indicates the magnitude of the stress on the insulation between the high-voltage winding H and the excitation winding E or the iron core F. The fully isolated winding may be earthed if necessary at either of the two terminals or at the centre tap, as shown; in the latter case, the output voltage will be symmetrical with respect to earth.

To generate voltages above a few hundred kV single-stage transformers according to Figure 1.1-1 are now rarely used; for economical and technical reasons one employs

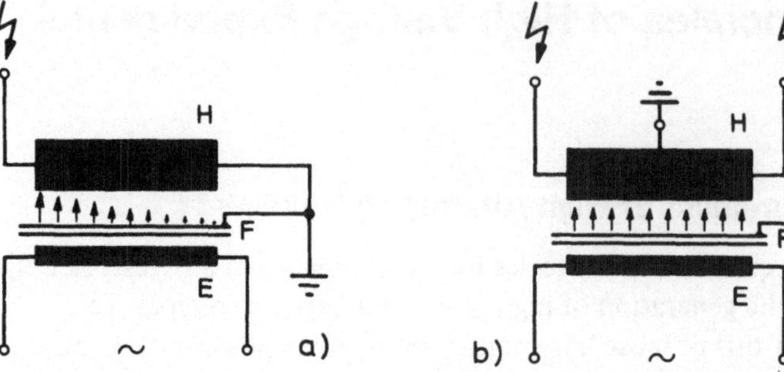

Fig. 1.1-1
Single stage test transformer circuits
E Excitation winding
H High-voltage winding
F Iron core
a) Single pole isolated
b) Fully isolated

instead a series connection of the high-voltage windings of several transformers. In such a cascade arrangement, the individual transformers must be insulated for voltages corresponding to those of the lower stages. The excitation windings of the transformers of all stages except the lowest will operate at high potential.

A frequently used circuit, introduced in 1915 by *W. Petersen, F. Dessauer* and *E. Welter*, is shown in Fig. 1.1-2. The excitation windings E of the upper stages are supplied from the coupling windings K of the stages immediately below. The individual stages, except the uppermost, must consist of three-winding transformers. When the temperature rise [*Grabner* 1967], the curve shape [*Matthes* 1959, *Müller* 1961] and the short-circuit voltage [*Pfestorf, Tayaram* 1960] are determined, it should be noted that the coupling and excitation windings of the lower stages have to transmit higher powers than those of the upper ones and accordingly have to be designed for higher loading. The magnitude of the power carried by the individual windings is indicated in Fig. 1.1-2 in terms of multiples of P.

The calculation of the total short-circuit impedance of a cascade arrangement from data for the individual stages will be demonstrated in Appendix 2. Test transformers in cascade connection have already been produced for voltages above 2 MV.

Fig. 1.1-2 Three-stage test transformer cascade
E Excitation winding
H High-voltage winding
K Coupling winding

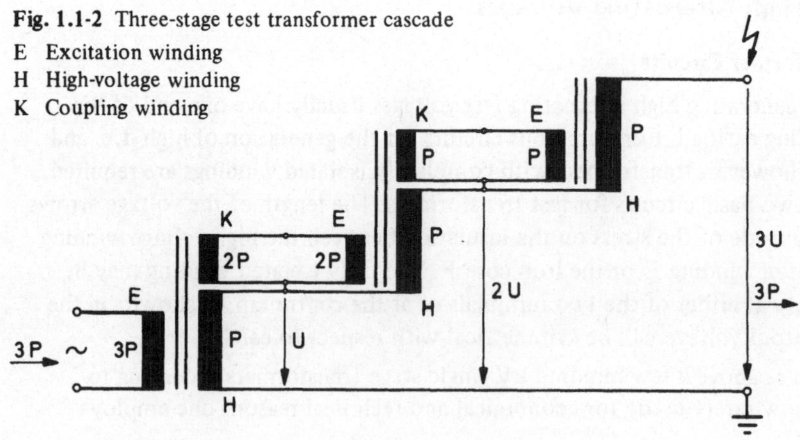

1.1.3 Construction of Test Transformers[1)]

For rating of not more than a few kVA, inductive voltage transformers can be used to generate high a.c. voltages. Low power test transformers are also similar in construction to voltage transformers with the same test voltage. For voltages up to about 100 kV epoxy resin insulation is widely used; oil-impregnated paper or oil with insulating barriers and spacers are found at higher voltages. At higher powers cooling of the windings becomes important, and the construction features resemble those of power transformers. Oil with barriers and oil-impregnated paper predominate as insulation.

Test transformers with cast resin insulation have at least their high-voltage winding moulded in epoxy resin. Fig. 1.1-3 shows a much simplified cross-section of such a transformer.

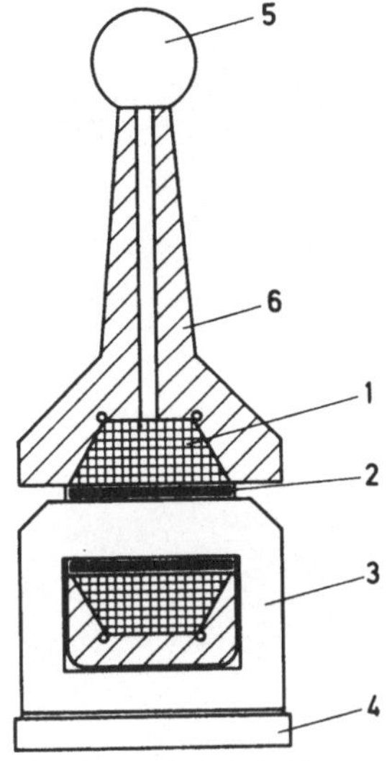

Fig. 1.1-3
Cross-section of a test transformer with cast resin insulation

1. High-voltage winding
2. Low-voltage winding
3. Iron core
4. Base
5. High-voltage terminal
6. Insulation

There are numerous designs for oil-insulated test transformers. In the tank type construction, shown in Fig. 1.1-4a, the active parts (core and windings) are enclosed in a metal container the surface of which provides useful self-cooling. However, at high working voltages the space requirement and high cost of the bushing is a disadvantage. In the insulated enclosure transformer type, as shown in Fig. 1.1-4b, the active parts are surrounded by an insulating cylinder. In general, transformers of this kind contain a relatively large quantity of oil and so have large thermal time constants in the case of overloading. Heat dissipation through the insulated enclosure is very small; consequently, closed-circuit cooling by means of external heat exchangers is necessary at high continuous rating. The advantage is that no bushings are required and that high-voltage electrodes with large radii of curvature can easily be fitted.

---

1) See also *Sirotinski* 1956; *Lesch* 1959; *Potthoff, Widmann* 1965; *Prinz* 1965; *Grabner* 1967.

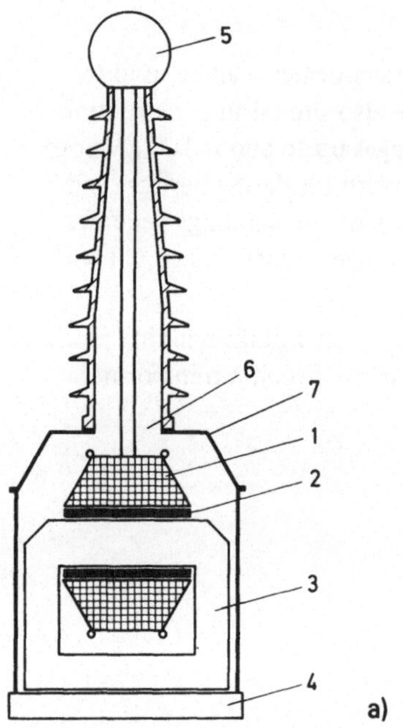

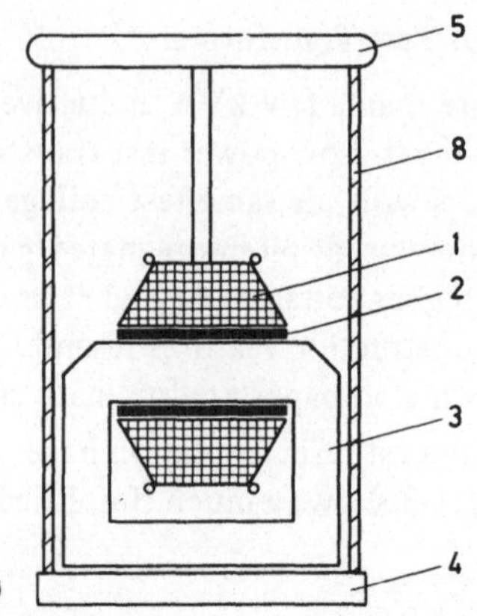

Fig. 1.1-4
Oil-insulated test transformers Nos. 1–5 as in Fig. 1.1-3
6 Bushing
7 Metal tank
8 Insulated enclosure
a) Tank design
b) Insulated enclosure

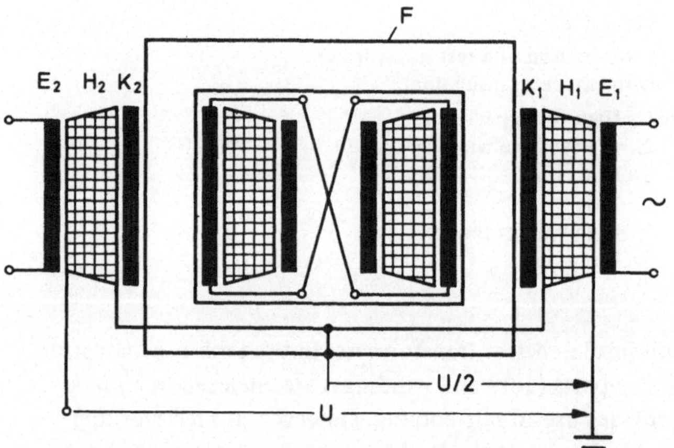

Fig. 1.1-5
A 2-stage cascade with common iron core at mid-potential
E_1, E_2 Excitation windings
H_1, H_2 High-voltage windings
K_1, K_2 Coupling windings
F Iron core

An advantageous and therefore frequently used arrangement of the active parts is shown in Fig. 1.1-5. It can be considered as a 2-stage cascade in which both stages have a common iron core F, which is at mid-potential and thus normally requires insulated mounting. For the symmetrical arrangement of the windings shown, E_1 or E_2 can be chosen for the primary excitation. If a cascade circuit is to be set up with a further transformer unit, the unenergized winding can be used as coupling for the next higher stage. For excitation via K_1, K_2, a symmetrical high voltage with respect to earth is obtained. The example of Fig. 1.1-5 shows the voltages to earth which occur when the right high-voltage terminal is earthed.

The described arrangement is especially advantageous at very high voltages and can be set up according to the tank type design with two bushings, and also according to the insulated enclosure type design. In the latter case however, the arrangement would be turned by 90° so that the two stages lie above each other.

1.1.4 Performance of Test Transformers

The working performance of test transformers is described only very inadequately with the aid of the usual transformer equivalent circuit; this is because the self-capacitance C_i of the high-voltage winding and the capacitance of the connected test object, the latter is a predominantly capacitive external load C_a, have considerable influence on the performance. On the other hand, the magnetization current can be neglected as long as there is no saturation of the iron core.

For approximate analysis of the working performance, the equivalent circuit shown in Fig. 1.1-6 is well suited. It comprises a series circuit of the short-circuit impedance $R_k + j\omega L_k$ and the total capacitance $C = C_i + C_a$ on the high-voltage side. $\widetilde{U}_1'$ is the secondary voltage due to transformation of the primary voltage $\widetilde{U}_1$. This equivalent circuit can also represent test transformers in cascade connection.

Since as a rule $R_k \ll \omega L_k$ and the secondary voltage $\widetilde{U}_2$ is then almost in phase with the primary voltage $\widetilde{U}_1$, we have:

$$U_2 \approx U_1' \cdot \frac{1}{1 - \omega^2 L_k C} \quad .$$

$(1 - \omega^2 L_k C)$ is always less than unity; thus series resonance leads to a capacitive enhancement of the secondary voltage. The amount of capacitive voltage enhancement can easily be calculated from the transformed short-circuit voltage u_k of the transformer, for the case when the capacitive load C just takes rated current I_n at rated voltage U_n and nominal frequency:

$$u_k = \frac{I_n \omega L_k}{U_n} = \omega^2 L_k C \quad .$$

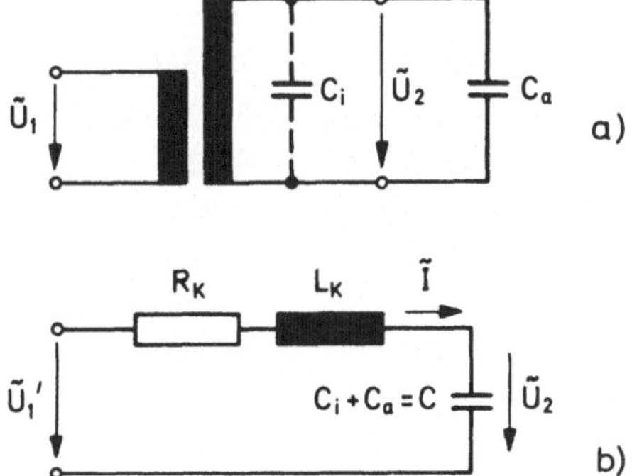

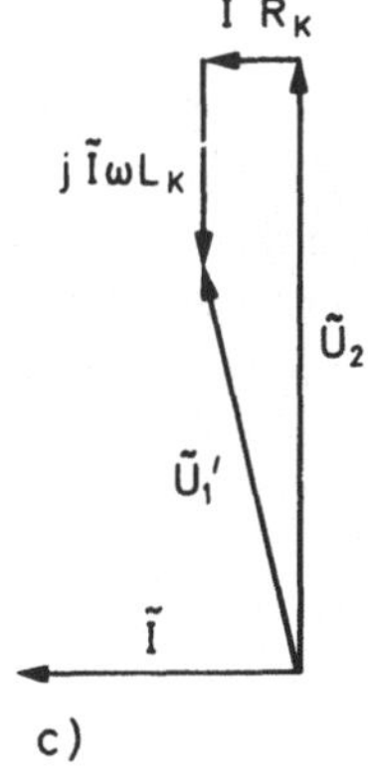

Fig. 1.1-6
Working performance of test transformers
a) Circuit diagram
b) Equivalent circuit
c) Phasor diagram

Thus a test transformer with u_K = 20 % will show a voltage enhancement of 25 % at nominal frequency when a capacitive load takes the rated current.

This voltage enhancement has to be taken into account, particularly for test transformers with high values of transformed short-circuit voltage, and above all when used at higher frequencies. There is then no longer a fixed ratio of primary to secondary voltage; for this reason determination of the value of the high-voltage output by voltage measurement on the low-voltage side of the transformer is inadmissible. This measurement would indicate values well below the real ones, and test object as well as test transformer could be endangered.

Test transformers, especially in cascade connection, represent spatially extended networks capable of oscillation. Harmonics of the primary voltage and the magnetization current may excite natural oscillations at various frequencies, and this can lead to considerable distortion of the secondary voltage.

Since the harmonics of the high voltage show pronounced dependence upon the load current and the value of the set voltage, one should take care that during specification tests the allowed deviation of the high voltage from the sinusoid of the same fundamental frequency does not exceed 5 % of the peak value. For this check, the peak value of the assumed ideal sinusoid may be chosen in such a way that the deviations of the actual shape, above and below the sinusoid, are a minimum (IEC-Publ. 60-2, 1973).

1.1.5 High Voltage Generation with Resonant Circuits

Fig. 1.1-6 was used to demonstrate the possibility of a very considerable voltage enhancement on the secondary side of a test transformer by series resonance with a capacitive load. This effect can be used for the generation of high alternating test voltages; to extend the tuning range, the short-circuit inductance of the test transformer is then augmented by a separate high-voltage inductor. The series resonant circuit formed by the inductance and the capacitance of the test object may be excited by a transformer of relatively low secondary voltage [*Kuffel, Abdullah* 1970].

Resonant circuits are particularly advantageous when the test object has a high capacitance, for instance, a high-voltage cable. The special advantage of such a circuit is that the output voltage deviates little from a sinusoid and that, due to the characteristics of the series resonant circuit, almost complete compensation of the reactive power required for the test object follows.

The Tesla transformer, named after its inventor, also belongs to the class of resonant circuits [*Marx* 1952; *Heise* 1964]. The circuit comprises a primary and a secondary oscillatory circuit in loose magnetic coupling. Periodic discharges of the primary capacitor via a spark gap will excite high frequency oscillations, typically in the frequency range 10^4 to 10^5 Hz. Depending upon the chosen circuit data and the transformation ratio of secondary to primary winding, voltages of more than 1 MV have been generated with Tesla transformers.

Measurement of High Alternating Voltages[1]

1.1.6 Peak Voltage Measurement with Sphere Gaps

Breakdown of a spark gap occurs within a few μs once the applied voltage exceeds the "static breakdown discharge voltage". Over such a short period the peak value of a power frequency voltage can be considered to be constant. Breakdown in gases will therefore always occur on the peak of low frequency a.c. voltages. With approximately homogeneous field gaps, for which the breakdown discharge times are particularly short, this behaviour is followed quite well to higher frequencies. Consequently the peak values of high a.c. voltages of frequencies up to about 500 kHz can be determined from the gap spacing at breakdown of measuring spark gaps in atmospheric air.

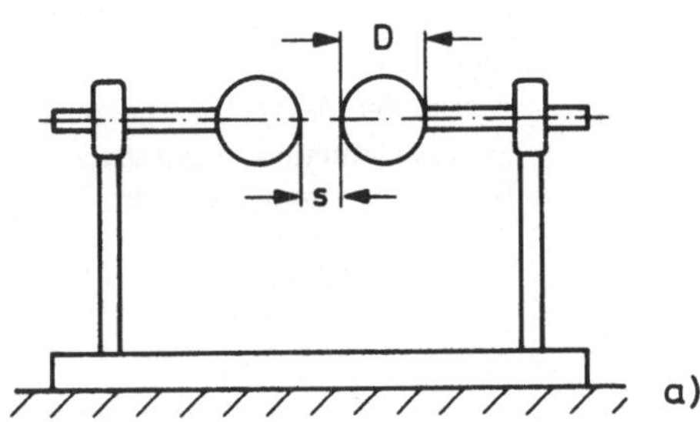

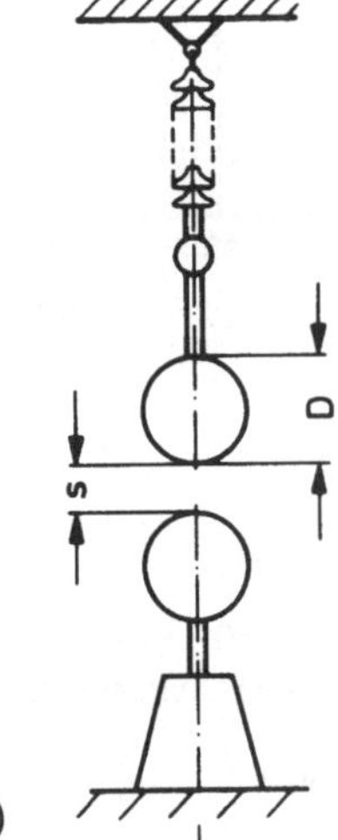

Fig. 1.1-7
Sphere gaps for voltage measurement
a) Horizontal arrangement
b) Vertical arrangement

Fig. 1.1-7 shows the two basic arrangements of sphere gaps for measuring purposes. The horizontal arrangement is usually preferred for sphere diameters $D < 50$ cm used for the lower voltage ranges; with the larger spheres the vertical arrangement is chosen; it is most suitable for measuring voltages with reference to earth potential.

The published specifications (IEC-Publ. 52-1960, BS 358) prescribe minimum clearances from objects disturbing the electric field and tabulate breakdown voltages for standard conditions ($b = 1013$ mbar, $\vartheta = 20$ °C) and various sphere diameters D as a function of the gap spacing s:

$$\hat{U}_{d_0} = f(D, s).$$

Humidity has no significant influence on the breakdown voltage of sphere gaps. Fig. 1.1-8 demonstrates the dependence of breakdown voltage upon gap spacing for various sphere

---

1) The techniques have been summarized by *Craggs, Meek* 1954; *Sirotinski* 1956; *Potthoff, Widmann* 1965; *Schwab* 1972; and others

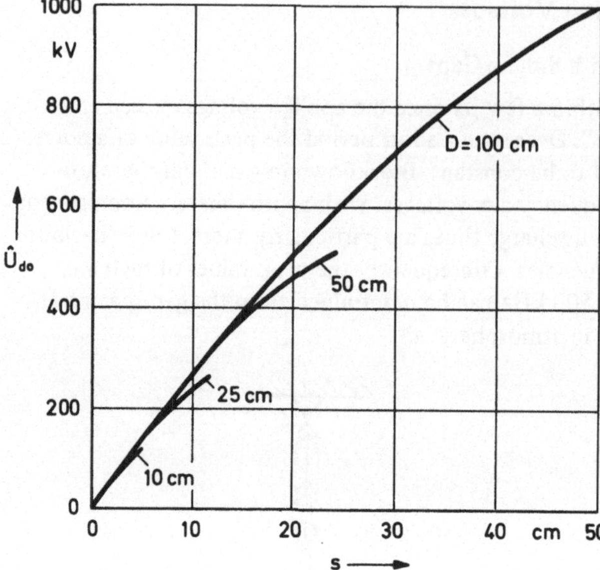

Fig. 1.1-8
Breakdown voltage U_{d_0} of sphere gaps as a function of gap spacing s, for various sphere diameters D

diameters. For measurements with sphere gaps, with increasing ratio s/D the field becomes increasingly inhomogeneous; at the same time the influence of the gap surroundings becomes greater, and so does the scatter in the values of breakdown voltages. Evidently the ratio s/D may not be too large. The minimum sphere diameter D for measurement of a voltage of amplitude $\hat{U}$ can be estimated from the following relationship:

$$D \text{ in mm} \geqslant \hat{U} \text{ in kV}.$$

It should be pointed out that during these measurements the tabulated values are only valid as long as the minimum clearances between the gap and the other parts of the setup are maintained.

Since the breakdown voltage $\hat{U}_d$ is proportional to the relative air density d in the range 0.9 ... 1.1, the actual breakdown voltage $\hat{U}_d$ at air density d may be found from the tabulated value $\hat{U}_{d_0}$ by applying the following formula:

$$\hat{U}_d \approx d\,\hat{U}_{d_0} = \frac{b}{1013}\,\frac{273+20}{273+\vartheta}\,\hat{U}_{d_0} = 0{,}289\,\frac{b}{273+\vartheta}\,\hat{U}_{d_0}$$

with b and ϑ in mbar[1]) and °C respectively.

Even under apparently ideal conditions, having made allowance for such factors as the air density, minimum clearances, smooth exactly spherical electrode surface and proper adjustment of the spacing, a measuring uncertainty of 3 % remains. Sphere gaps are now rarely used for measuring voltages above 1 MV, because they require excessive space and

1) 1 mbar = 100 N/m^2 ≈ 0.75 Torr

are expensive. Continuous voltage measurement is obviously impossible with sphere gaps, since the voltage source is short-circuited at the instant of measurement. The method can however be used to record and check measured points, for instance, in the point-to-point recording of a calibration curve showing the dependence of the high voltage on the primary voltage of a transformer for a certain test setup. Inspite of their disadvantages, sphere gaps can be useful and versatile devices in a high-voltage laboratory. Apart from voltage measurement, they can also be used as voltage limiters, as voltage-dependent switches, as pulse sharpening gaps and as variable high-voltage capacitors, etc.

1.1.7 Peak Voltage Measurement Using Measuring Capacitors

The circuit shown in Fig. 1.1-9, suggested by *Chubb* and *Fortescue* in 1913, is well suited for exact and continuous measurement of the peak value of a high a.c. voltage against earth.

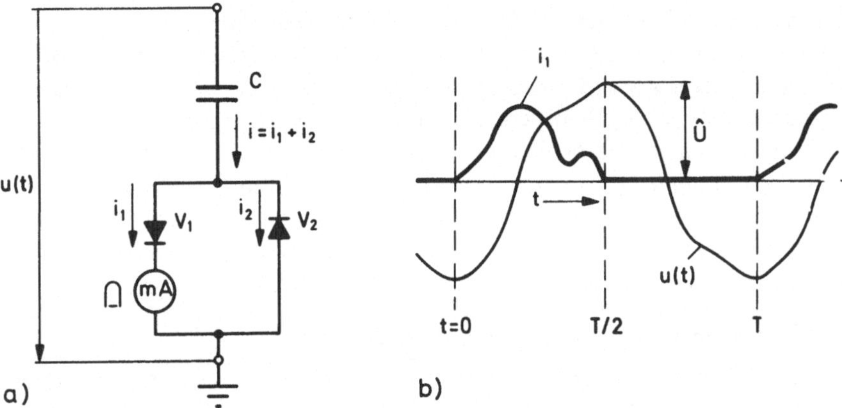

Fig. 1.1-9 Peak voltage measurement according to *Chubb* and *Fortescue*
a) Circuit b) Current and voltage curves

A charging current i, given by the rate of change of the applied voltage u(t) to be measured, flows through the high-voltage capacitor C and is passed through two antiparallel rectifiers V_1 and V_2 to earth. The arithmetic mean value $\overline{I}_1$ of the current i_1 in the left-hand branch is measured with a moving-coil instrument; as shown below, provided that certain conditions are fulfilled, this current is proportional to the peak value $\hat{U}$ of the high voltage.

If the behaviour of the rectifiers is assumed ideal, then for the conducting period of V_1 we have:

$$i_1 = i = C\frac{du}{dt} \quad \text{for } t = 0 \ldots T/2$$

$$\overline{I}_1 = \frac{1}{T}\int_0^T i_1\,dt = \frac{1}{T}\int_{u(0)}^{u(T/2)} C\,du = \frac{C}{T}\left[u\left(\frac{T}{2}\right) - u(0)\right].$$

If the voltage is symmetrical with reference to the zero line:

$$u\left(\frac{T}{2}\right) - u(0) = 2\hat{U}$$

and with T = 1/f, we obtain:

$$\hat{U} = \overline{I}_1 \frac{1}{2\,fC}.$$

If a circuit with full-period rectification (Graetz circuit) is used instead of the half-period rectifier circuit shown in the figure, the factor 2 in the denominator of the above equation should be replaced by 4. For the derivation of this expression, it was not assumed that u(t) is a sinusoid, though when passive rectifiers (especially semiconductor diodes) are used, we have to demand that the high voltage to be measured does not have more than one maximum per half-period. The use of synchronous mechanical rectifiers or controllable rectifiers (oscillating contacts, rotating rectifiers) allows correct measurement of alternating voltages with more than one maximum per half-period. Oscillographic monitoring of the high-voltage shape is necessary and is usually done by observing the current i_1, which may have one crossover only in each half-period. As the frequency f, the measuring capacitance C and the current $\overline{I}_1$ can be determined precisely, measurement of symmetrical a.c. voltages using the technique of *Chubb* and *Fortescue* with the appropriate outlay is very accurate, and is suitable for the calibration of other peak voltage measuring devices [*Boeck* 1963]. A disadvantage for technical routine measurements is the dependence of the reading upon the frequency and the need to monitor the curve.

1.1.8 Peak Value Measurement with Capacitive Voltage Dividers

Several rectifier circuits have been developed which permit the measurement of peak values of high a.c. voltages with the aid of capacitive dividers. Compared with the circuit of *Chubb* and *Fortescue,* most of these methods have the advantage that the reading is practically independent of frequency, and multiple extrema per half-period of the voltage to be measured can be permitted.

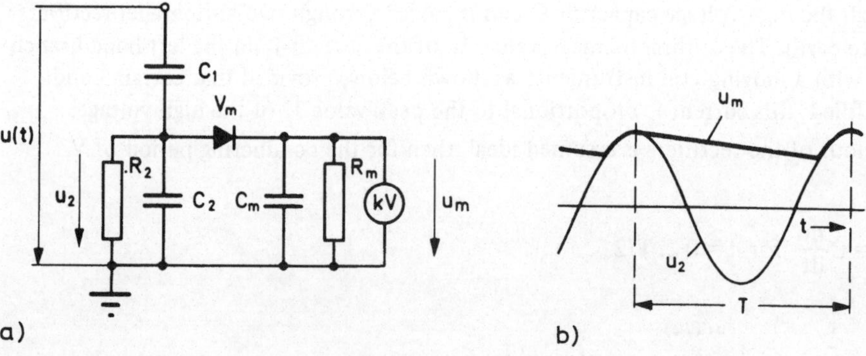

Fig. 1.1-10 Peak voltage measurement with a capacitive divider
a) Circuit b) General form of the voltage

The half-period circuit shown in Fig. 1.1-10 is particularly simple and also sufficiently accurate for most purposes. In this circuit, the measuring capacitor C_m is charged to the peak value $\hat{U}_2$ of the lower arm voltage $u_2(t)$ of a capacitive divider. The resistor R_m which discharges the capacitor C_m is necessary to ensure an adequate response to reductions in the applied voltage. The choice of time constant for this discharge process is determined by the desired response of the measuring arrangement, whereby the internal resistance of the connected measuring instrument must be taken into account. In general, one chooses:

$$R_m C_m < 1 \text{ s}.$$

On the other hand, the time constant must be large compared with the period $T = 1/f$ of the alternating voltage to be measured, so that the voltage u_m across C_m does not drop significantly between recharging cycles; the time dependence $u_m(t)$ is indicated in Fig. 1.1-10b. The appropriate condition here is:

$$R_m C_m \gg \frac{1}{f}.$$

The resistance R_2 parallel to C_2 is necessary in order to prevent charging of C_2 by the current flowing through the rectifier V_m. The value of R_2 must be chosen in such a way that the direct voltage drop across R_2 which causes d.c. charging of C_2 remains as small as possible, thus we must have:

$$R_2 \ll R_m,$$

on the other hand, the capacitive divider ratio should be affected little by R_2, and so:

$$R_e \gg 1/(\omega C_2).$$

Provided all these conditions can be satisfied, the relation between the peak value of the high voltage and the indicated voltage $\hat{U}_m$ is given by:

$$\hat{U} = \frac{C_1 + C_2}{C_1} \hat{U}_m.$$

The indicating instrument should have a high input impedance; electrostatic voltmeters, high sensitivity moving-coil instruments and resistance or electrometer amplifiers with analog or digital indication are suitable. Measuring range changes are usually effected by changing C_2.

The postulates made above for the relative values of the circuit components are not quite compatible and limit the obtainable accuracy, particularly at low frequencies. The properties can be improved with more elaborate circuitry [*Zaengl, Völcker* 1961].

The overall achievable accuracy, however, not only depends upon the properties of the low-voltage measuring circuit, but also upon those of the high-voltage capacitor. Measuring capacitors for very high voltages are frequently not fully screened, so that additional errors due to stray fields may be expected [*Lührmann* 1970].

1.1.9 Measurement of r.m.s. Values by Means of Electrostatic Voltmeters

When a voltage u(t) is applied to an electrode arrangement, such as the one shown in Fig. 1.1-11a for example, the electric field produces a force F(t) which tends to reduce the spacing s of the electrodes. This attractive force can be calculated from the change of energy of the electric field:

$$W(t) = \frac{1}{2} C u^2(t).$$

The capacitance C of the arrangement depends on the spacing s.

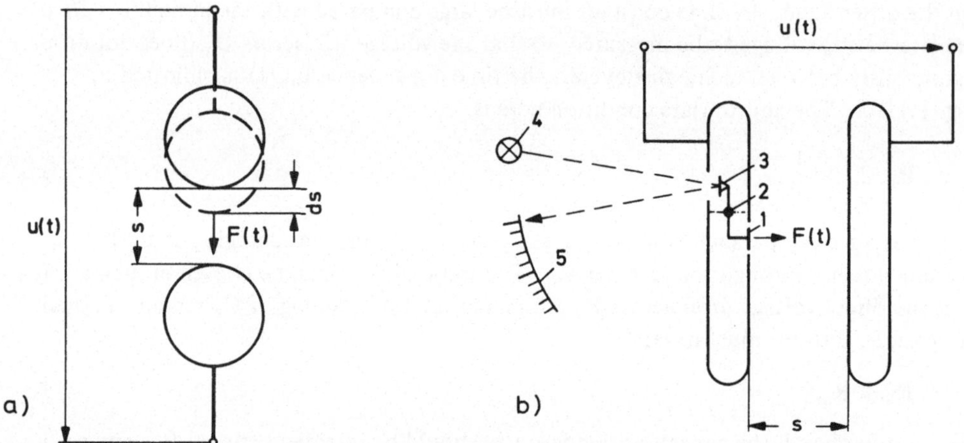

Fig. 1.1-11 Electrostatic voltmeters for high voltages
a) Using spherical electrodes (after Hueter)
b) Using a movable electrode segment (after Starke and Schröder)

1. Movable electrode segment
2. Axis of rotation
3. Mirror
4. Light source
5. Scale

The time-dependent force is obtained from the law of conservation of energy dW + F ds = 0 assuming disconnection of the voltage source [*Küpfmüller* 1965]. Taking into account that the charge Cu(t) is independent of s, it follows:

$$F(t) = -\frac{dW(t)}{ds} = \frac{1}{2} u^2(t) \frac{dC}{ds}.$$

If the arithmetic mean value $\overline{F}$ of the force is calculated from this expression, the linear relationship between $\overline{F}$ and the square of the r.m.s. value of the applied voltage is apparent:

$$\overline{F} = \frac{1}{2} \frac{dC}{ds} \frac{1}{T} \int_0^T u^2(t)\, dt \sim U_{rms}^2.$$

The influence of the factor dC/ds depends upon the way in which the force $\overline{F}$ is translated into an indication. In general, dC/ds changes over the measuring range, so that the deflection no longer shows strict quadratic dependence.

As an example of an electrostatic measuring device, the design of *Starke* and *Schroeder* is shown in a simplified form in Fig. 1.1-11b. The force F(t) acts on a small plate 1 mounted on a cranked lever with a central pivot; at the other end of the lever is a small mirror 3 which deflects a light beam for the optical indication. The taut band suspension 2 provides the restoring torque.

Electrostatic voltmeters are characterized by their very high internal resistance and very small capacitance; they are thus also useful for the direct measurement of high-frequency high voltages extending to the MHz region.

1.1.10 Measurements with Voltage Transformers

High alternating voltages can be measured extremely accurately with voltage transformers. Although these devices are widely used in power supply networks, they are rarely used in laboratories for measurements of voltages above 100 kV.

The basic circuits of single pole isolated inductive and capacitive voltage transformers, for the measurement of voltages with respect to earth, are shown in Fig. 1.1-12.

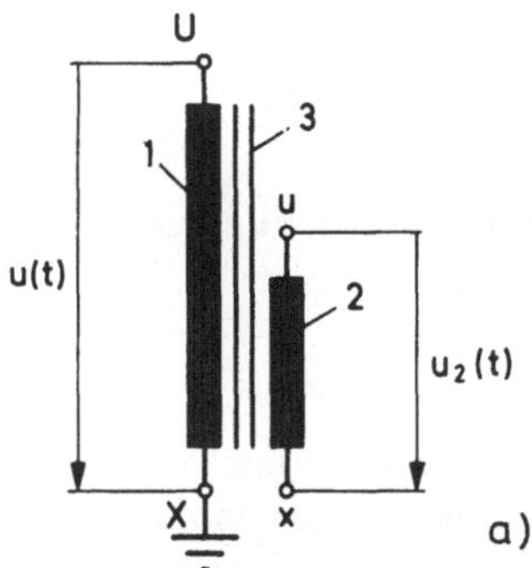

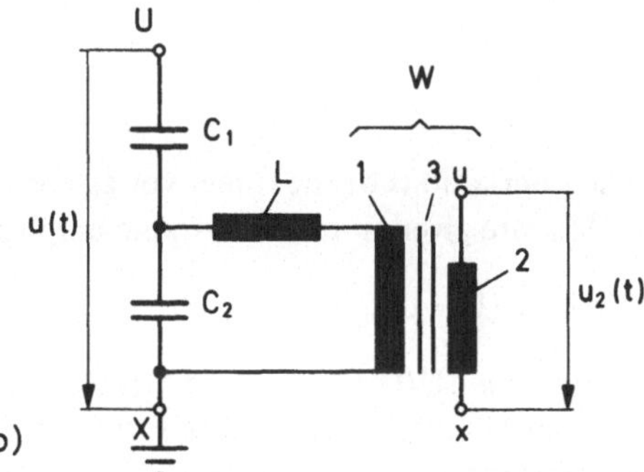

Fig. 1.1-12 Basic circuits of voltage transformers

a) Inductive voltage transformers
1 Primary winding
2 Secondary winding
3 Iron core

b) Capacitive voltage transformer
C_1, C_2 Divider capacitors
L Resonance inductor
W Matching transformer (marking as under a)

Inductive voltage transformers for very high voltages can be built only at great expense, since, for the comparatively low test frequency of 50 Hz, the product of magnetic flux and number of turns of the high-voltage winding, by the laws of induction, takes very large values. This leads to expensive designs.

The type of capacitive voltage transformer used extensively in supply networks is often considered unsuitable for normal testing work, mainly because it imposes a high capacitive load upon the voltage source.

Inductive and capacitive voltage transformers will thus be used in laboratory practice only when particularly precise measurement of moderate voltages is required. The secondary voltage of a voltage transformer will reproduce the shape of the primary voltage, irrespective of the secondary load. Depending upon the type of measuring device connected, it is possible to measure the peak value, the r.m.s. value or the high-voltage curve.

1.2 Generation and Measurement of High Direct Voltages

There are numerous applications for high direct voltages in the laboratory, such as insulation testing of arrangements with high capacitance, e.g. capacitors or cables, and fundamental investigations in discharge physics and dielectric behaviour. Technical uses include the generation of X-rays, precipitators, paint spraying and powder coating.
The most common generation methods of high direct voltages employ rectification of high alternating voltages, often using voltage multiplication; electrostatic generators are also in use. The high direct voltages are usually measured by means of high resistance measuring resistors or by electrostatic voltmeters.

1.2.1 Characteristic Parameters of High Direct Voltages

The d.c. test voltage is defined as the arithmetic mean value [IEC Publ. 60-2, 1973]:

$$\bar{U} = \frac{1}{T}\int_0^T u(t)\,dt.$$

Periodic fluctuations of the direct voltage between the peak value $\hat{U}$ and the minimum value U_{min} are given in terms of ripple amplitude:

$$\delta U = \tfrac{1}{2}\,(\hat{U} - U_{min}).$$

The expression $\delta U/\bar{U}$ is called the "ripple factor". Sometimes, with regard to the working mode of certain measuring methods, the r.m.s. value U_{rms} as in section 1.1.1 is quoted. For a well-smoothed direct voltage, $\delta U/\bar{U} \ll 1$ and we have:

$$\bar{U} \approx \hat{U} \approx U_{rms}.$$

Generation of High Direct Voltages[1]

1.2.2 Properties of High-Voltage Rectifiers

Rectifiers in laboratory circuits for the generation of high direct voltages are usually series-connected stacks of semiconductor diodes or high vacuum valves (Fig. 1.2-1).

[1] A comprehensive review of methods may be found in *Craggs, Meek* 1954; *Sirotinski* 1956; *Lesch* 1959; *Kuffel, Abdullah* 1970, and others

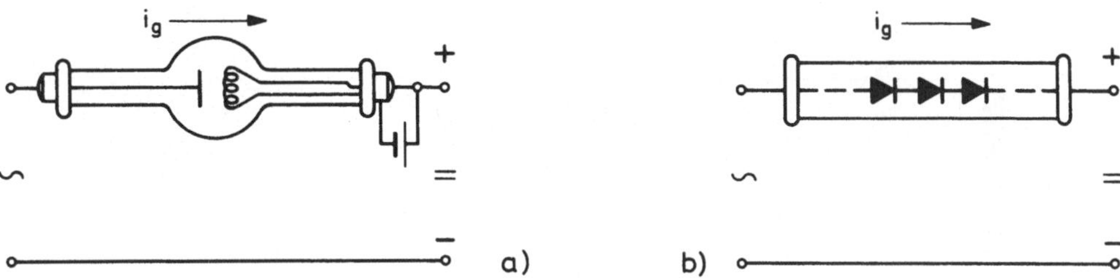

Fig. 1.2-1 High-voltage rectifiers

a) High vacuum rectifier b) Semiconductor rectifier

In high vacuum rectifiers the current is carried by electrons emitted from a thermionic cathode and accelerated towards the anode by the electric field. These rectifiers are available for use with peak inverse voltages of up to 100 kV. Although in laboratory practice high vacuum rectifiers have been replaced by semiconductor rectifiers, which are more convenient to use as no provision need be made for cathode heating, the former are of great advantage in X-ray installations since they can function as X-ray tubes at the same time.

In contrast to high vacuum rectifiers, semiconductor diodes are not true valves since they allow a small but finite current flow in the blocked condition. The following guiding values may be given for inverse voltages and forward currents of the more commonly used semiconductor rectifiers:

Semiconductor material	Selenium	Germanium	Silicon
Peak inverse voltage per element (V)	30–50	150–300	1000–2000
Loading capacity of the depletion layer (A/cm^2)	0.1–0.5	50–150	50–150

Compared with silicon diodes, selenium rectifiers are more bulky and have a lower efficiency. Laboratory applications, however, require currents of the order of only a few hundred mA. Selenium rectifiers are very suitable here, since, because of the high capacitance of the depletion layer, they can be used in multi-element stacks with peak inverse voltages up to about 600 kV without any further voltage grading capacitors.

1.2.3 The Half-Period Rectifier Circuit

The simplest circuit for the generation of a high direct voltage is the half-period rectification shown in Fig. 1.2-2. A load R is supplied from a high-voltage transformer T via a rectifier V. We assume that the secondary voltage $u_T(t)$ of the transformer is a sinusoid and the rectifier is ideal, i.e. with zero forward resistance and zero reverse current. Figs. 1.2-2b and 1.2-2c show the shape of the voltage across the load for steady-state conditions for circuits with and without a smoothing capacitor C, indicated by a dashed line.

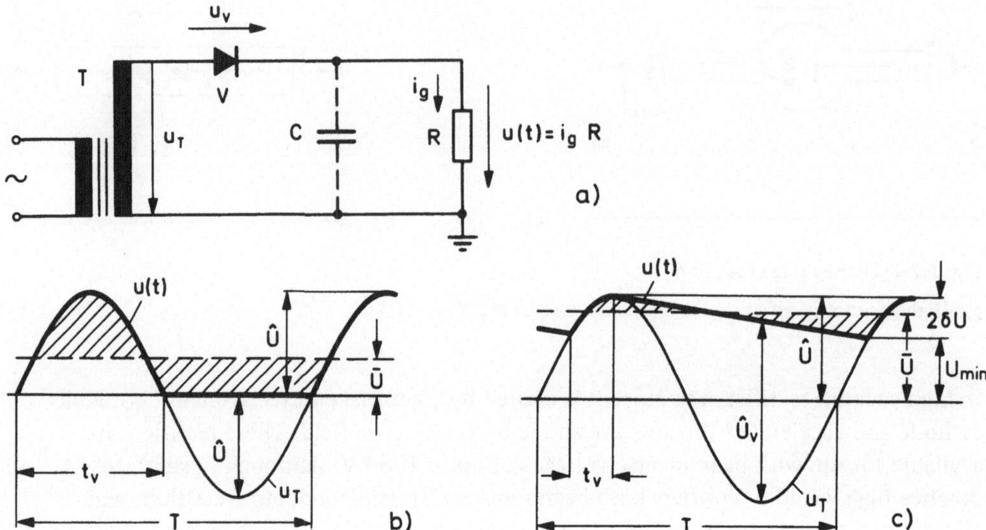

Fig. 1.2-2 Half-period rectification with ideal circuit elements

a) Circuit

b) Output voltage curve without smoothing capacitor C

c) Output voltage curve with smoothing capacitor C

The circuit without the smoothing capacitor C will give a pulsating direct voltage with the following characteristic values:

$$\hat{U} = \hat{U}_T; \quad \bar{U} = \frac{1}{\pi}\,\hat{U}; \quad U_{rms} = \frac{1}{2}\,\hat{U}.$$

The conducting period t_v of the rectifier is equal to half the period T of the alternating voltage. The peak inverse voltage across the rectifier in the reverse direction is:

$$\hat{U}_V = \hat{U}_T.$$

For the circuit with the smoothing capacitor C a smoothed direct voltage with residual ripple is obtained. We have:

$$\hat{U} = \hat{U}_T; \quad \bar{U} \approx \hat{U} - \delta U.$$

The better the smoothing of the voltage, the shorter the current flow period t_V will be. Thus, during the conducting period of the rectifier only a short forward current pulse flows each time, the peak inverse voltage being:

$$\hat{U}_V \approx 2\,\hat{U}_T.$$

Referring to Fig. 1.2-2c, the ripple value can easily be calculated. When

$$t_v \ll T = \frac{1}{f} \quad \text{and} \quad \delta U \ll \bar{U}$$

then the exponential discharge curve for the capacitor C during the blocking period of V can be replaced by a straight line. From the change of charge on the smoothing capacitor during the blocking period, we have:

$$2\delta UC \approx \int_0^T i_g \, dt = T\bar{I}_g$$

$$\delta U \approx \bar{I}_g \frac{1}{2\,fC} \, .$$

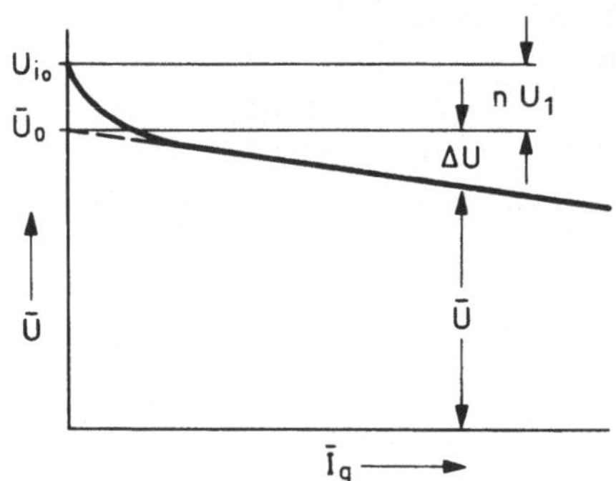

Fig. 1.2-3
Load characteristic for semiconductor rectifiers

In full-period rectification the time intervals between successive recharging, and thus the ripple δU, are reduced to one half.

The usual methods for reducing ripple in rectifier circuits are increasing the size of the smoothing capacitor, the frequency and the number of phases. In laboratory setups frequencies up to some 10 kHz are frequently used and ripple factors of a few percent are common. With multi-stage filters and/or electronic ripple compensation, ripple factors of the order of 10^{-5} can be obtained.

In the design of circuits, the non-ideal behaviour of rectifiers has to be taken into account; in particular the forward voltage drop for current flow in the conducting direction must be allowed for. A non-linear relationship between the direct current $\bar{I}_g$ and the direct voltage $\bar{U}$ is the consequence. Fig. 1.2-3 shows the typical load characteristic for a semiconductor rectifier. For $\bar{I}_g = 0$, the ideal no-load voltage U_{i_0} is the peak voltage of the transformer, $U_{i_0} = \hat{U}_T$. Linear extrapolation of the load characteristic for high currents to $\bar{I}_g = 0$ gives an intercept with the ordinate at a value $\bar{U}_0$ which is lower than U_{i_0} by the value U_1 for each element, practically independent of the current. The output voltage of a rectifier of n elements in series can therefore, except for very low currents, be described by the equation:

$$\bar{U} = \bar{U}_0 - \Delta U = (U_{i_0} - nU_1) - k\bar{I}_g,$$

where k is a proportionality factor depending upon the type of rectifier; the voltage U_1 is of the order of 0.6 ... 1.2 V.

1.2.4 Voltage Multiplier Circuits

The most widely used multiplier circuits will now be described, assuming idealized elements. A common property of all the circuits considered here is that they are only able to supply relatively low currents and are therefore not suitable for high current applications, such as high-voltage direct current transmission. The voltage curves are shown to illustrate the principles of the various circuits. For simplification the excitation windings of the high-voltage transformers T have been omitted in the circuit diagrams.

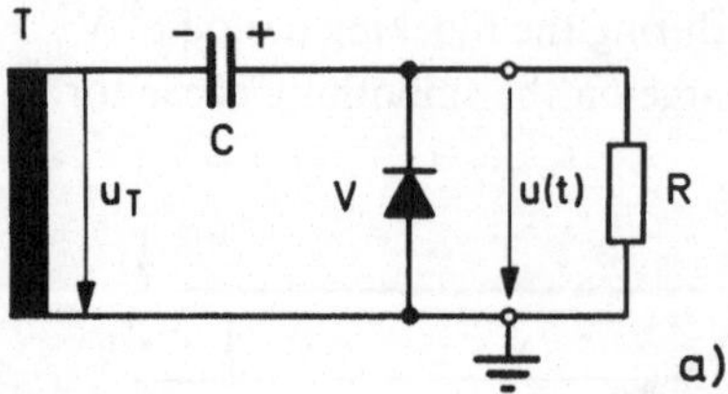

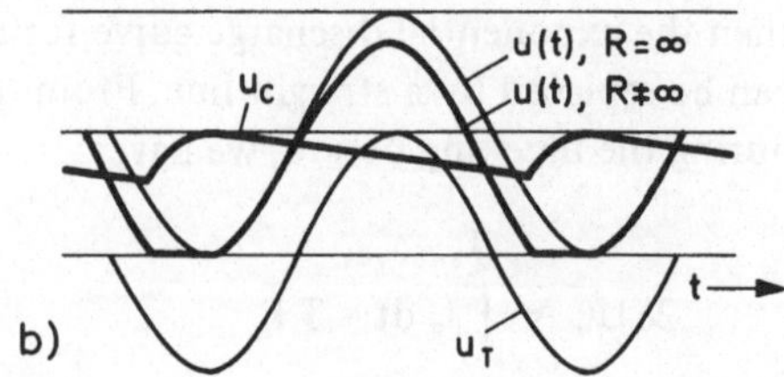

Fig. 1.2-4 Villard circuit a) Circuit diagram b) Voltage curve

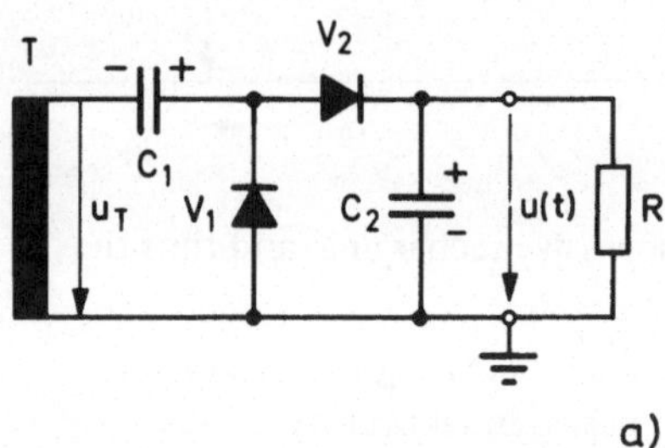

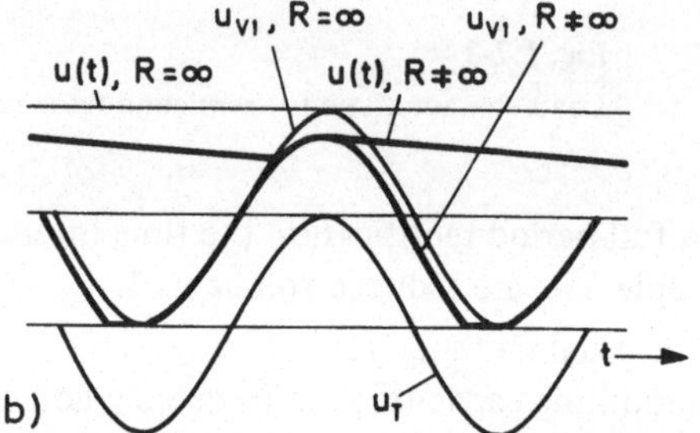

Fig. 1.2-5 Greinacher doubler-circuit a) Circuit diagram b) Voltage curve

Villard Circuit: This circuit, shown in Fig. 1.2-4, is the simplest doubling circuit. The blocking capacitor C is charged to the peak value $\hat{U}_T$ and thus increases the potential of the high-voltage output terminal with respect to the transformer voltage by this amount. For no-load conditions, we have:

$$\bar{U} = \hat{U}_T; \quad \hat{U} = 2\hat{U}_T; \quad \hat{U}_V = 2\hat{U}_T.$$

Smoothing of the output voltage u(t) is not possible.

Greinacher Doubler-Circuit: Fig. 1.2-5 shows the extension of the Villard circuit by a rectifier V_2, which enables the smoothing capacitor C_2 to be connected. For no-load conditions, we have:

$$\bar{U} = \hat{U} = 2\hat{U}_T; \quad \hat{U}_{V1} = \hat{U}_{V2} = 2\hat{U}_T.$$

The sum of the peak inverse voltages of the rectifiers in this circuit is twice the output voltage $\bar{U}$. This is true of any rectifier circuit providing a smooth direct voltage.

Zimmermann-Wittka Circuit: If two Villard circuits are connected in opposition as in Fig. 1.2-6, an unsmoothed direct voltage is produced between the output terminals, with a peak value three times that of the transformer voltage and, under no-load conditions, a mean output voltage $\bar{U} = 2\hat{U}_T$. This circuit may be earthed at any point, provided the transformer has an adequate winding insulation; this also applies to the other circuits described above.

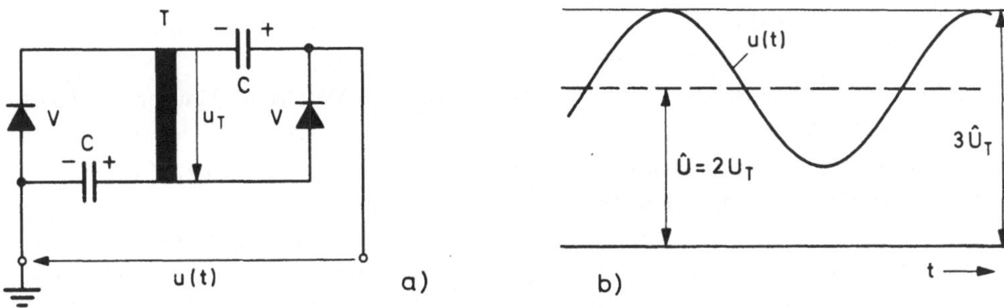

Fig. 1.2-6 Zimmermann-Wittka circuit (no-load condition) a) Circuit diagram b) Voltage curve

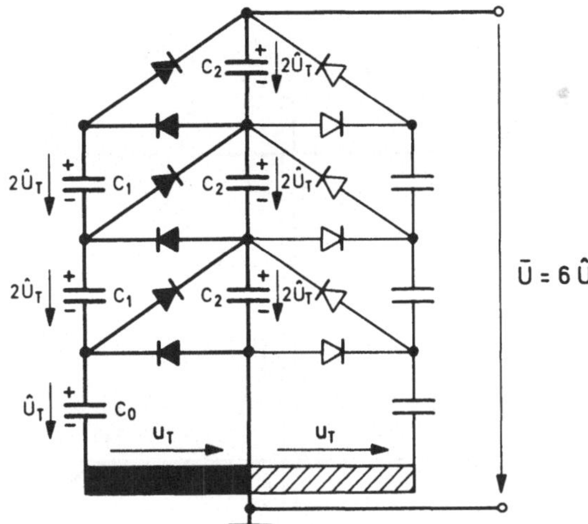

Fig. 1.2-7
Greinacher cascade circuit (no-load condition)

Greinacher Cascade Circuit: This circuit, suggested in 1920 by *H. Greinacher* and which is also known as the *Cockroft-Walton* multiplier, is the most important method for the generation of very high direct voltages. It is an extension of the Greinacher doubler-circuit.

A three-stage circuit is shown in Fig. 1.2-7 as an example; many practical circuits comprise only the parts shown in bold lines. The voltages indicated apply for ideal circuit elements and the no-load condition. To achieve a more uniform voltage drop, it proves useful to choose the capacitance C_0 of the lowest unit to be twice C_1, that of the capacitors above it. The capacitor stack comprising the elements C_2 in series serves, among other functions, as a smoothing capacitor. The load characteristics of a cascade can be derived from those of the half-period rectifier circuit of Fig. 1.2-2.

The simple unsymmetrical multi-stage circuit shows relatively high ripple δU and voltage drop ΔU when loaded, even at increased input frequency; this becomes worse as the number of stages is increased. Extending the circuit to a symmetrical Greinacher cascade, by including the thinly drawn lines shown in Fig. 1.2-7, reduces ripple and voltage drop and thus offers very considerable advantages [*Baldinger* 1959].

Greinacher cascades have been constructed for voltages of 5 MV; typical current capabilities of test setups are of the order of 10 mA.

Casading of separate rectifier circuits: When relatively high output currents, of the order of 100 mA or more are required, a series connection of separate rectifier circuits as shown in Fig. 1.2-8 is an economical solution. In this way low ripple and voltage drops can be achieved even when output currents are high. The alternating current inputs to the individual circuits must be provided at the appropriate high potential; this can be done by means of isolating transformers or by means of individual alternators driven via insulating shafts.

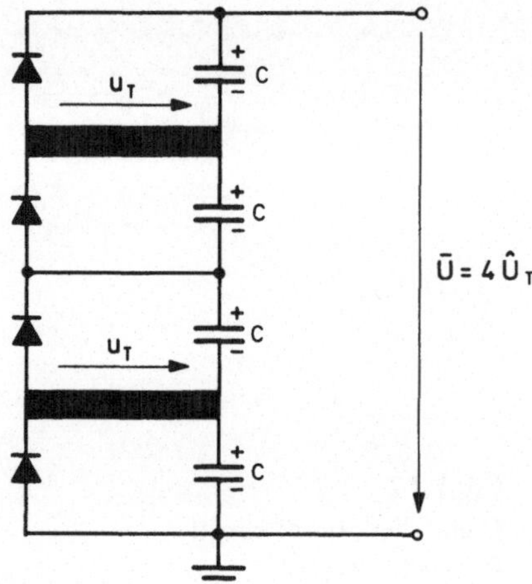

Fig. 1.2-8 Example of cascade rectifier circuits (no-load condition)

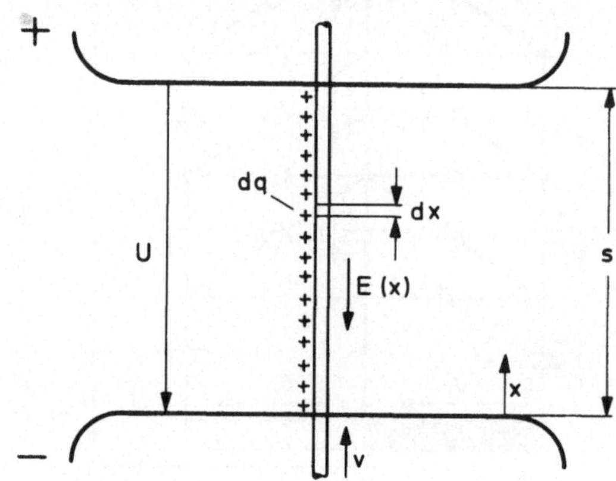

Fig. 1.2-9 To illustrate the working principle of electrostatic generators

1.2.5 Electrostatic Generators

In electromagnetic generators, current carrying conductors are moved against the electromagnetic forces acting upon them. In electrostatic generators, movement of electrically charged particles occurs against the electrostatic forces acting upon them.

The working principle of an electrostatic generator will be explained using Fig. 1.2-9 as illustration. An insulated belt of width b with charge carrier density σ is suspended in the electric field E(x) between two electrodes with spacing s. The charge on a strip of height dx is given by:

$$dq = \sigma\, b\, dx.$$

The force acting upon the entire belt is:

$$F = \int_0^s dF = \int_0^s E(x)\,dq = \sigma b \int_0^s E(x)\,dx.$$

If the belt is moved with constant velocity v = dx/dt against this force, the necessary mechanical power is:

$$P = Fv = \sigma b v \int_0^s E(x)\,dx.$$

Since

$$I = \frac{dq}{dt} = \sigma b v \quad \text{and} \quad U = \int_0^s E(x)\,dx$$

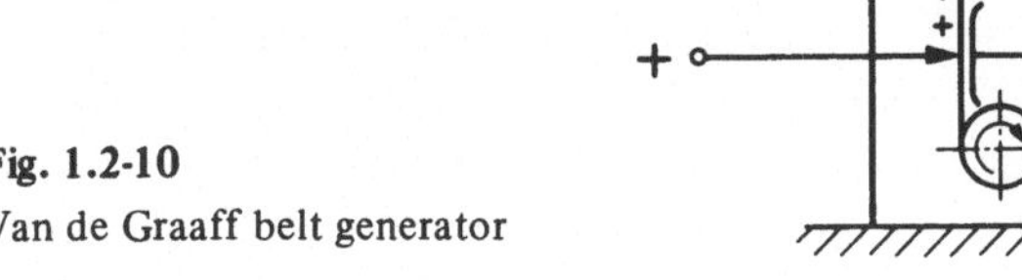

Fig. 1.2-10
Van de Graaff belt generator

it can be seen that the mechanical power required for the drive is equal to the electrical power output IU.

The most common type of electrostatic generator is the belt generator devised in 1931 by *R. I. van de Graaff,* the working principle of which is illustrated in Fig. 1.2-10. An insulated belt is run over rollers and electrostatically charged by an excitation arrangement. A strongly inhomogeneous electrode configuration is used for this purpose; charge carriers formed by collision ionisation at the sharp electrode are trapped by the belt on their way to the opposite electrode. A similar arrangement at the high-voltage end serves to discharge the belt. If another excitation arrangement of the opposite polarity is provided at the high-voltage level for the downwards moving side of the belt, twice the amount of current is obtained.

Belt generators of the pressure tank type have already been constructed for voltages above 10 MV, where currents less than 1 mA are usual [*Herb* 1959]. Instead of the insulated belt one can also use highly insulating liquids or dust-like solid materials as carriers of the electric charge.

For voltages up to some 100 kV diverse types of electrostatic machines with drum or disc-shaped rotors have been built. Among some of the advantages of these machines are good control over the constancy of the output voltage, as well as low self-capacitance, which results in an essentially safe high-voltage device [*Felici* 1957].

Measurement of High Direct Voltages [1])

1.2.6 Measurements with High-Voltage Resistors

The measurement of a d.c. voltage can, with the aid of resistors, be reduced to the measurement of a direct current. The basically very simple circuit is shown in Fig. 1.2-11.

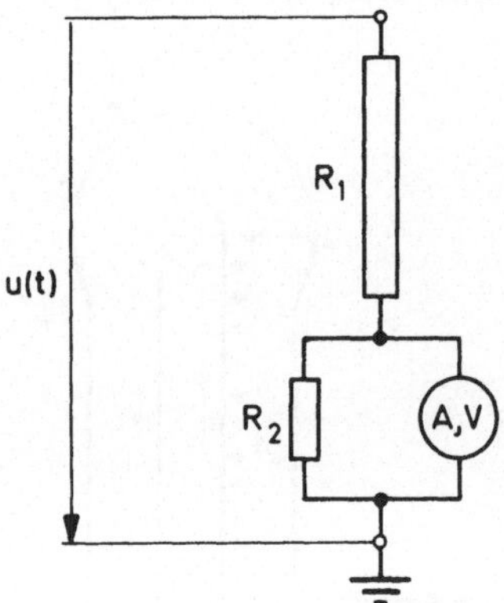

Fig. 1.2-11
Measurement of a direct voltage by means of a series resistor or a resistive divider

For application to high voltages there is the problem that the measuring current must be chosen to be very small, of the order of 1 mA for example, because of the permitted loading of the voltage source and heating of the measuring resistor. A small current is however easily falsified by error currents; these occur in the form of leakage currents in insulating materials and on insulating surfaces, and also as a result of corona discharges. Certain details of the design of high-voltage measuring resistors shall be given in section 2.4.1.
The characteristic parameter of the d.c. voltage measured depends on the working principle of the ammeter at earth potential, and connected in series with the measuring resistor. A sensitive moving-coil instrument is usually chosen, the indication of which is a measure of the arithmetic mean value $\bar{U}$ of the d.c. voltage. The measuring range is easily changed in every case by parallel connection of a resistance R_2 to the measuring instrument, which turns the series resistor into a resistive voltage divider. Instead of the ammeter, a voltmeter with an internal resistance preferably much larger than R_2 may also be connected.

1.2.7 Measurement of r.m.s. Values by Means of Electrostatic Voltmeters

As may be seen from the description, in section 1.1.4, of the working principle of electrostatic voltmeters, this type of instrument can also be used for direct voltages. Electrostatic voltmeters do in fact represent the best way of measuring high d.c. voltages directly. It is a question of a loss-free measurement which can also be performed when no current may be drawn from the voltage source.
In this method, voltage measurement is reduced to measurement of a field strength at an electrode, which is particularly illustrated by the arrangement indicated in Fig. 1.1-11b.

[1]) Comprehensive treatment in *Böning* 1953; *Craggs, Meek* 1954; *Sirotinski* 1956; *Paasche* 1957; *Kuffel, Abdullah* 1970; *Schwab* 1972, and others.

For high direct voltages, space charges will occur when electrodes of small radius of curvature are used and the system is not fully screened. These space charges, or surface charges adhering to the surface of insulating materials, can affect the field strength at the rotating electrode segment and so result in considerable error.

1.2.8 Voltmeter and Field Strength Meter Based upon the Generator Principle

Consider the electrode arrangement shown in Fig. 1.2-12a, where a measuring electrode of area A, assumed to be at earth potential, has constant surface charge density $\epsilon_0 E$ produced by the steady field strength E. The total charge on the measuring electrode is given by:

$$q = \int_{(A)} \epsilon_0 E \, dA = \epsilon_0 A E.$$

The charge q is now allowed to vary between the values q_{max} and q_{min} as is shown in Fig. 1.2-12b, this being done by periodic covering and uncovering of a portion of the measuring electrode by an earthed plate. An alternating current $i(t) = dq/dt$ then flows in the earth lead; the curves of the positive and negative half-periods are the same if the covering and unçovering movement is uniform. The arithmetic mean value of the current between two crossovers is then:

$$\frac{1}{T/2} \int_0^{T/2} \frac{dq}{dt} \, dt = \frac{2}{T} (q_{max} - q_{min}).$$

For rectification, this value corresponds to the arithmetic mean value $\overline{I}$ taken over a whole period.

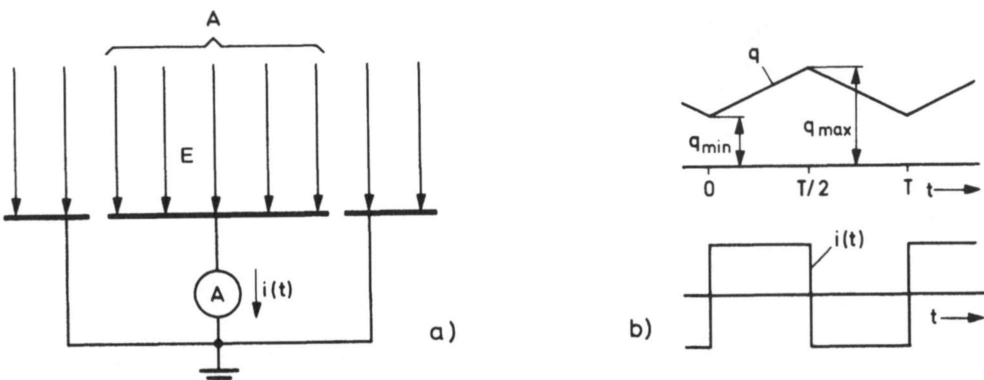

Fig. 1.2-12 Measurement of voltage and field strength according to the generator principle

a) Schematic measuring arrangement

b) Charge and current curves

If the measuring electrode is completely covered at $t = 0$, q_{min} would be zero, and we have:

$$\bar{I} = \frac{2}{T} q_{max} = \frac{2}{T} \epsilon_0 AE.$$

Thus, $\bar{I}$ is proportional to the field strength and can be used to measure the latter. If the frequency of the mechanical movement is high, even low steady field strengths can be measured well because of the correspondingly high dq/dt. This principle was indeed applied for the first time in 1926, by *A. Matthias* and *H. Schwenkhagen* in thunderstorm investigations, to measure electric field strengths at ground level. A different type of field strength meter, instead of covering the electrode, uses an oscillatory movement of the measuring electrode in the field direction to generate the alternating current i(t).
Using the arrangement shown schematically in Fig. 1.2-13 as an example, we shall show how a direct voltage U may be measured according to the same principle [*Kind* 1956]. The two measuring electrodes 1 and 1′ are alternately passed underneath the semicircular opening 2 of the earthed plate 3 by the drive; this produces a partial capacitance, varying

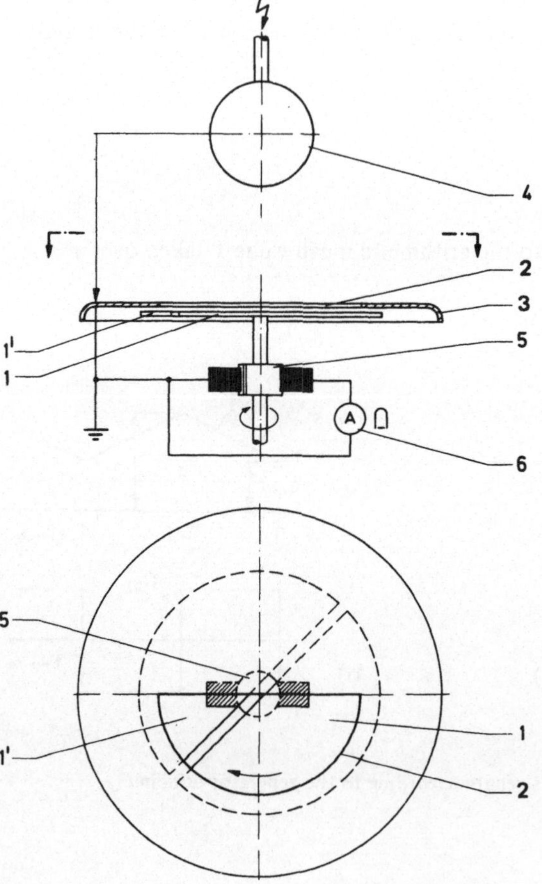

Fig. 1.2-13
Voltmeter with the sphere-plate electrode configuration

1,1′ Revolving semicircular discs
2 Semicircular opening
3 Earthed covering plate
4 High-voltage electrode
5 Commutator
6 Ammeter

between zero and a maximum value, between each of the measuring electrodes and the high-voltage electrode 4. At constant rate of revolution, therefore, a periodic alternating current i(t) flows between the measuring electrodes, which is rectified by the commutator 5. The arithmetic mean value $\overline{I}$ after rectification can be recorded by the moving-coil ammeter 6. Owing to the proportionality of the field strength E at the electrodes and the voltage U to be measured, $\overline{I}$ is proportional to U. If, in order to determine the surface charge, we introduce the maximum value C_m of the periodically varying partial capacitance between one measuring electrode and the high-voltage electrode, we have:

$$q_{max} = C_m U \quad \text{and} \quad q_{min} = 0$$

and it follows that:

$$\overline{I} = \frac{2}{T} C_m U.$$

The principle just described has been applied in various different ways [*Prinz* 1939; *Schwab* 1972]. It is of particular significance to the measurement of voltages and field strengths at high direct voltages, since the measurement is a no-load one.

1.2.9 Other Methods for the Measurement of High Direct Voltages

The method of measuring alternating voltages using sphere gaps, described in 1.1.6, is also suitable for the determination of the peak value $\hat{U}$ of high direct voltages. Rod gaps as in 2.4.3 may also be chosen instead of spheres if necessary.

Fundamentally different methods for the measurement of high direct voltages have been developed for special cases of application in physics. Those methods which allow the measured quantity to be expressed in terms of base units and of accurately known fundamental constants, are of particular scientific significance. For example, to calibrate the voltage measuring devices of elementary particle accelerators, protons are accelerated in an electric field which is proportional to the voltage to be measured. At certain kinetic energies of these protons, on collision with light atomic nuclei, resonant nuclear transformations occur which permit very exact determination of the applied d.c. voltage [*Jiggins, Bevan* 1966].

1.2.10 Measurement of Ripple Voltages

Ripple voltages are alternating voltages which diverge greatly from the sinusoidal form; they may therefore be represented by a Fourier series. For smoothed d.c. voltages, the peak values δU of the ripple voltages are always much smaller than $\overline{U}$, which is why an oscilloscopic measurement, performed on a resistive divider, for example, is too insensitive. Hence, circuits are used which enable the ripple time dependence $u(t) - \overline{U}$ to be measured directly.

Fig. 1.2-14 shows a simple circuit in which a high-voltage capacitor C separates the ripple from the d.c. voltage. The voltage divider made up of C and R has a divider ratio of zero for d.c. voltages; for an alternating voltage with angular frequency ω on the other hand, we have:

$$\frac{\tilde{U}_2}{\tilde{U}} = \frac{jR\omega C}{1 + jR\omega C}.$$

Now if the condition

$$u_2(t) \approx u(t) - \bar{U}$$

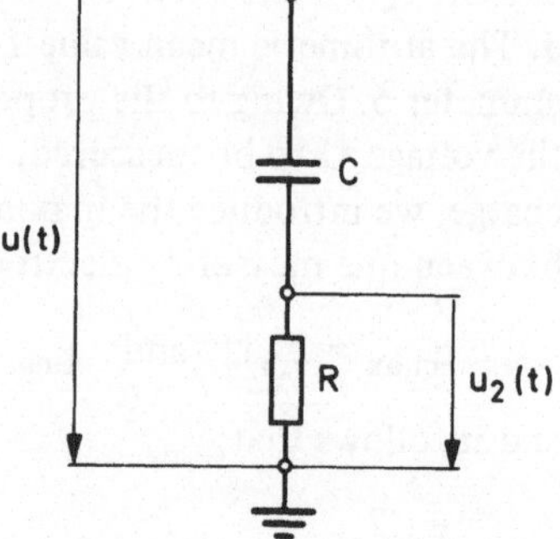

Fig. 1.2-14
Circuit for measuring ripple voltages

is to be well satisfied, the divider ratio must be as near to 1 as possible for all frequencies in the ripple spectrum, which is the case when

$$R\omega C \gg 1.$$

This is easily fulfilled for the basic frequency ω and so faithful reproduction of the ripples is assured.

1.3 Generation and Measurement of Impulse Voltages

Impulse voltages are required in high-voltage tests to simulate the stresses due to external and internal over-voltages, and also for fundamental investigations of the breakdown mechanisms. They are usually generated by discharging high-voltage capacitors through switching gaps onto a network of resistors and capacitors, whereby voltage multiplier circuits are often used. The peak value of impulse voltages can be determined with the aid of measuring gaps, or better, be measured by electronic circuits combined with voltage dividers. The most important measuring device for impulse voltages is, however, the cathode-ray oscilloscope, which allows the complete time characteristic of the voltage to be determined by means of voltage dividers; occasionally analog digital converter equipment may be used instead of the oscilloscope.

1.3.1 Characteristic Parameters of Impulse Voltages

In high-voltage technology a single, unipolar voltage pulse is termed an impulse voltage; three important examples are shown in Fig. 1.3-1, with reference to the possible characteristic parameters. The time dependence, as well as the duration of the impulse voltage, depend upon the method of generation. For basic experiments, rectangular impulse voltages are often used which rise abruptly to an almost constant value, as well as wedge-shaped impulse voltages characterized by a rise which is as linear as possible up to breakdown, and described simply by the steepness S. For testing purposes, double exponential impulse voltages have been standardized; without appreciable oscillation these rapidly reach a maximum, the peak value $\hat{U}$, and finally drop less abruptly to zero. If an inten-

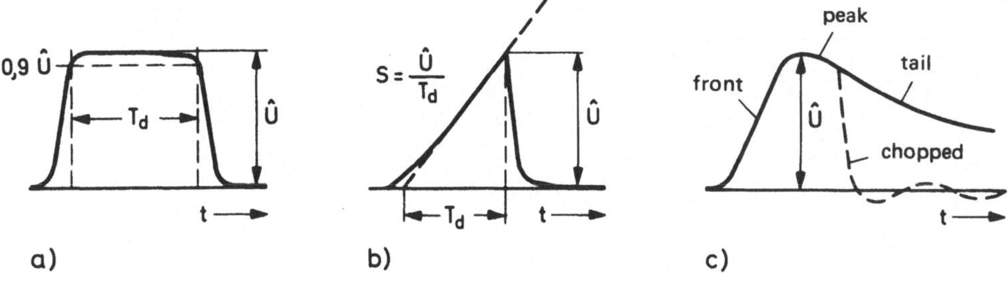

Fig. 1.3-1 Examples of impulse voltages a) Rectangular impulse voltage b) Wedge-shaped impulse voltage c) Double exponential impulse voltage

tional or unintentional breakdown occurs in the high-voltage circuit during the impulse, leading to a sudden collapse of the voltage, this is then called a chopped impulse voltage. The chopping can occur on the front, at the peak or in the tail section of the impulse voltage. The transient phenomenon thereby induced is mainly responsible for the oscillations indicated in Fig. 1.3-1c.

For overvoltages following lightning strokes, the time required to reach the peak value is of the order of 1 μs; they are named atmospheric or external overvoltages. Voltages generated in a laboratory to simulate these are called lightning impulse voltages. For internal overvoltages, occurring as a consequence of switching operations in high-voltage networks, the time taken to reach the peak value is at least about 100 μs. Their reproduction in the laboratory is effected by switching impulse voltages; these are of approximately the same shape as lightning impulse voltages, but last considerably longer.

In the case of impulse voltages for testing purposes the shape of the voltage is determined by certain time parameters for the fron and tail, as shown in Fig. 1.3-2 (IEC Publ. 60-2 (1973)). Since the true shape of the front of lightning impulse voltages is often difficult to measure, the straight line $0_1 S_1$ through the points A and B is introduced as an auxiliary construction on the front, to characterize the latter. Then the time T_s to front, as well as the time T_r to half-value, being the time from 0_1 to the point C, are also determined. In general, lightning impulse voltages of shape 1.2/50 are used, which means an impulse voltage with $T_s = 1.2\ \mu s \pm 30\ \%$ and $T_r = 50\ \mu s \pm 20\ \%$. On the other hand, recording the much slower switching impulse voltage presents no difficulties; hence the true origin 0 and the true peak S can be utilized for standardization. For tests with switching impulse voltages the shape 250/2500 is often used, which corresponds to $T_{cr} = 250\ \mu s \pm 20\ \%$ and $T_h = 2500\ \mu s \pm 60\ \%$ (T_{cr} = time to crest, T_h = time to half value)[1]. To denote the duration of the switching impulse voltage, the time T_d during which the instantaneous value of the voltage lies above 0.9 Û is also often given instead of T_h.

[1]) IEC Publication 60-2 (1973): High-Voltage Test Techniques, Test Procedures

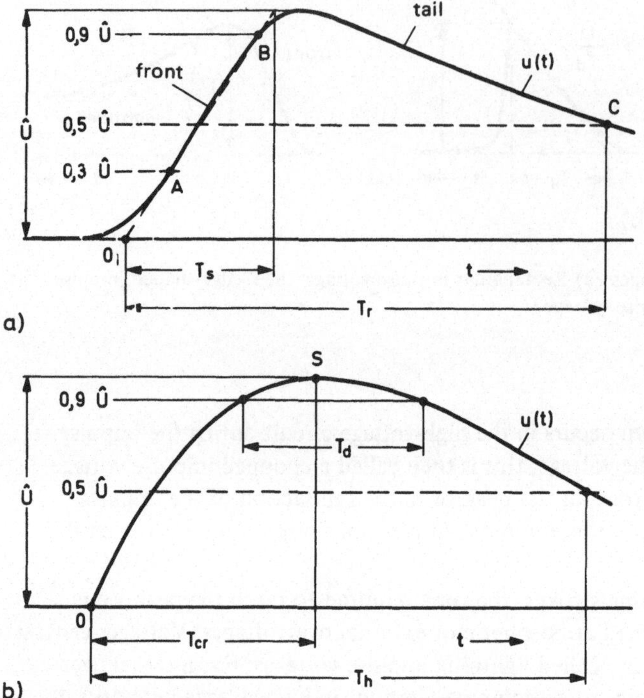

Fig. 1.3-2
Characteristic parameters of standard test impulse voltages
a) Lightning impulse voltage
b) Switching impulse voltage

The curves of lightning impulse voltages often have high-frequency oscillations superimposed, the amplitude of which may not exceed 0.05 Û in the region of the peak. It is assumed in this case that the frequency of the oscillations is at least 0.5 MHz, otherwise the actually observed maximum value of the voltage is taken as the peak value of the lightning impulse voltage.

Generation of Impulse Voltages

1.3.2 Capacitive Circuits for Impulse Voltage Generation[1]

Fig. 1.3-3 shows the two most important basic circuits, denoted "circuit a" and "circuit b", used for the generation of impulse voltages. The impulse capacitor C_s is charged via a high charging resistance to the direct voltage U_0 and then discharged by ignition of the switching gap F. The desired impulse voltage u(t) appears across the load capacitor C_b. The circuits a and b differ from one another in that, in the one case, the discharge resistor R_e is connected in front of, and in the other, behind the damping resistor R_d.

The value of the circuit elements determines the curve shape of the impulse voltage. The basic working principle of both circuits can be readily understood from the following

[1]) Comprehensive treatment in *Craggs, Meek* 1954; *Strigel* 1955; *Widmann* 1962; *Helmchen* 1963, and others

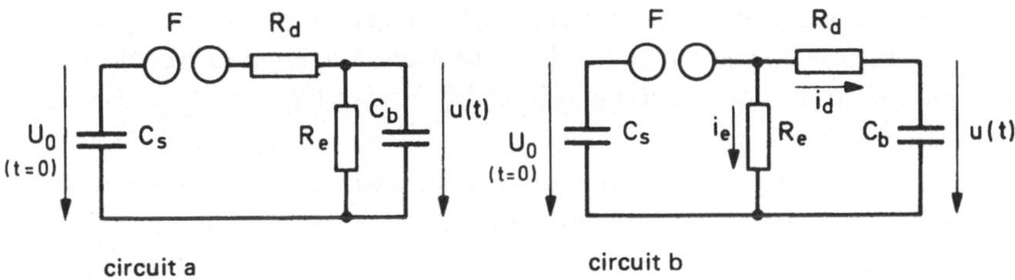

Fig. 1.3-3 Basic diagrams of impulse voltage circuits

simple considerations. The short time to front requires rapid charging of C_b to the peak value $\hat{U}$, and the long time to tail, slow discharge. This is achieved by $R_e \gg R_d$. Immediately after ignition of F at $t = 0$, almost the full charging voltage U_0 appears across the series combination of R_d and C_b in both circuits. The smaller the value of the expression $R_d C_b$, the faster is the rate at which the voltage u(t) reaches its peak value. The peak value $\hat{U}$ cannot be greater than is determined by distribution of the initially available charge $U_0 C_s$ onto $C_s + C_b$. For the utilization factor η therefore we have:

$$\eta = \frac{\hat{U}}{U_0} \leqslant \frac{C_s}{C_s + C_b} \, .$$

Since for a given charging voltage $\hat{U}$ should generally be as high as possible, one will choose $C_s \gg C_b$. The exponential decay of the impulse voltage on the tail would then, in circuit a, occur with the time constant $C_s (R_d + R_e)$, and in circuit b with the time constant $C_s R_e$. The impulse energy transformed during a discharge is then:

$$W = \frac{1}{2} C_s U_0^2 \, .$$

If the highest possible charging voltage is substituted for U_0 in this expression, we obtain the maximum impulse energy as an important characteristic parameter of the impulse voltage generator.

In the above explanation of the operating mode of the circuits, it was assumed that at $t = 0$ the impulse capacitors C_s were charged to a voltage U_0. U_0 is the value of the charging voltage at which F breaks down, either by itself or by means of an auxiliary discharge. Thus, for self-triggered operation, an increase in the peak value of the impulse voltage $\hat{U}$ can only be achieved by increasing the spacing of F. Merely increasing the direct voltage applied in front of the charging resistor would only result in C_s charging up faster to the value U_0, and F breaking down spontaneously in shorter intervals of time. Hence the impulse rate would increase and not the amplitude of the impulse voltage generated.

For given d.c. charging voltage, to obtain impulse voltages with as high a peak value as possible, the multiplier circuit proposed by *E. Marx* in 1923 is commonly used. Several identical impulse capacitors are charged in parallel and then discharged in series, obtaining in this way a multiplied total charging voltage, corresponding to the number of stages.

The mechanism of the Marx circuit will be explained with the aid of the impulse generator shown in Fig. 1.3-4, with n = 3 stages in circuit b connection. The impulse capacitors of the stages C_s' are charged to the stage charging voltage U_0', via the high charging resistors R_L' in parallel.

When all the switching gaps F break down, the capacitors C_s' will be connected in series, so that C_b is charged via the series connection of all the damping resistors R_d'; finally, all C_s' and C_b will discharge again via the resistors R_e' and R_d'. It is expedient to choose $R_L' \gg R_e'$. The n-stage circuit can be reduced to a single stage equivalent circuit, such as circuit b, where the following relationships are valid:

$$U_0 = n U_0' \qquad R_d = n R_d'$$

$$C_s = \frac{1}{n} C_s' \qquad R_e = n R_e'$$

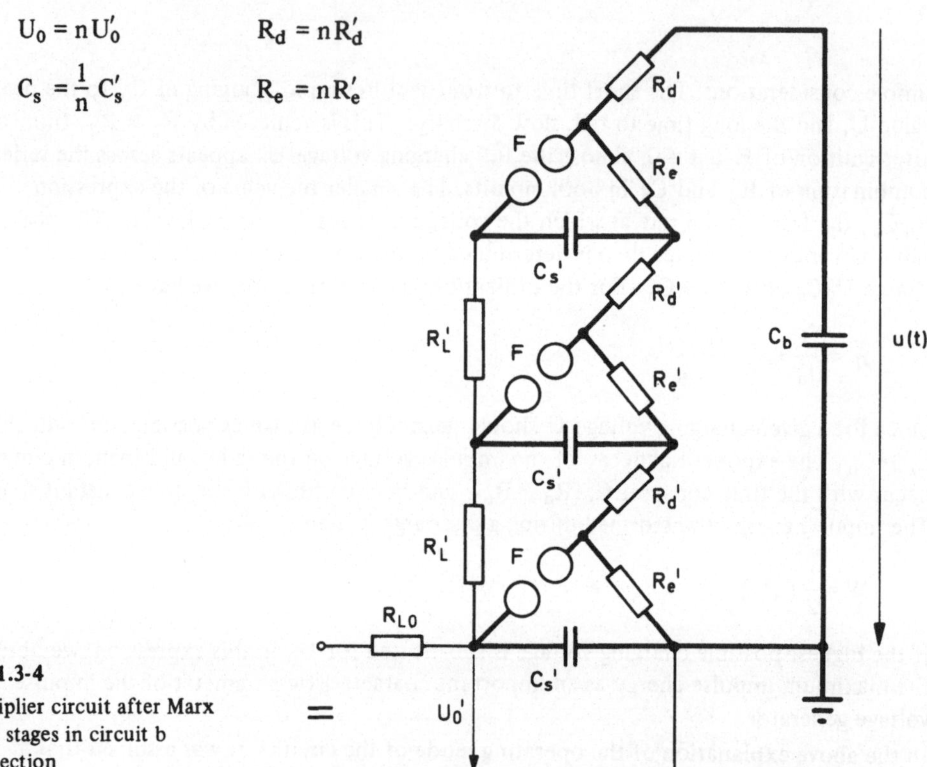

Fig. 1.3-4
Multiplier circuit after Marx for 3 stages in circuit b connection

If, in the circuit shown in Fig. 1.3-4, the discharge resistors R_e' in each stage are connected in parallel to the series combination of R_d', F and C_s', an impulse generator of type circuit a is obtained.

For the operation of the Marx circuit it is essential that all the switching gaps F, which are normally sphere gaps of adjustable spacing, break down almost simultaneously. This is usually achieved by setting the lowest sphere gap to a slightly smaller spacing or triggering it first by an auxiliary discharge. As a result of the unavoidable earth capacitance of the upper stages, transient over-voltages appear across the spheres at the higher levels, causing their breakdown as well [*Rodewald* 1969; *Heilbronner* 1971]. This breakdown mechanism forbids the distribution of C_b between the individual stages, since the transient phenomena essential for the breakdown of all the gaps would then be prevented. In the case of large generators, and particularly if switching impulse voltages are also to be generated, it could be worthwhile to use mechanical switches instead of charging resistors and to time-trigger more of the switching gaps F with an auxiliary discharge instead of only the lowest.

Impulse voltage generators have already been built for voltages of the order of a few MV and for impulse energies of a few hundred kWs, where the charging voltages per stage are usually of the order of 100 ... 300 kV. The utilization factor η depends on the shape of the impulse voltage to be generated and generally lies between 0.6 and 0.9. It is also principally higher for circuit b than for circuit a, especially for impulse voltages with comparatively shorter times to tail.

1.3.3 Calculation of Single-Stage Impulse Voltage Circuits

For the design of impulse voltage circuits it is necessary to establish relationships between the values of the circuit elements and the characteristics of the voltage shape. Because of the higher utilization factor, impulse generators are built predominantly in the basic circuit b connection. For this reason the impulse shape has been calculated in Appendix 3 for this circuit, using the symbols shown in Fig. 1.3-3b.

For the impulse voltage curve the solution is:

$$u(t) = \frac{U_0}{R_d C_b} \frac{T_1 T_2}{T_1 - T_2} \left(e^{-t/T_1} - e^{-t/T_2}\right).$$

It is seen that the impulse voltage is given by the difference of two exponentially decaying functions with time constants T_1 and T_2. Fig. 1.3-5 shows the curve which reaches the peak value $\hat{U}$ at time T_{cr}.

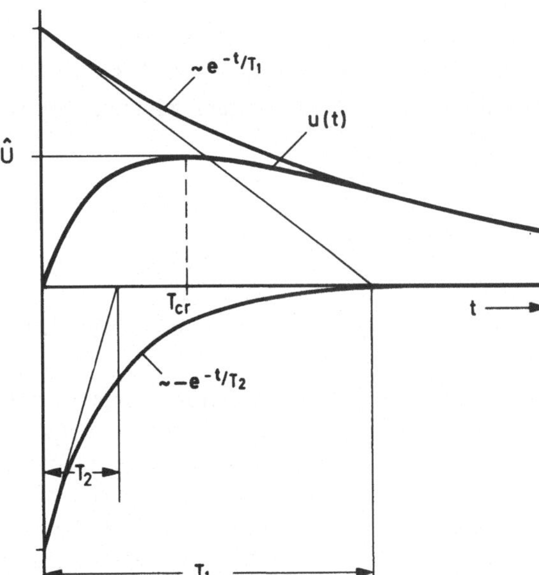

Fig. 1.3-5
Calculation of double exponential impulse voltages

With the usually satisfied approximation

$$R_e C_s \gg R_d C_b$$

the following simple expressions are obtained for circuit b:

$$T_1 \approx R_e(C_s + C_b), \qquad T_2 \approx R_d \frac{C_s C_b}{C_s + C_b}, \qquad \eta \approx \frac{C_s}{C_s + C_b}.$$

For circuit a of Fig. 1.3-3a the same general solution is true, but with:

$$T_1 \approx (R_d + R_e)(C_s + C_b), \qquad T_2 \approx \frac{R_d R_e}{R_d + R_e} \frac{C_s C_b}{C_s + C_b},$$

$$\eta \approx \frac{R_e}{R_d + R_e} \frac{C_s}{C_s + C_b}$$

The impulse shape is described uniquely by T_1 and T_2. Consequently the characteristics as in Fig. 1.3-2 must also be functions of T_1 and T_2. Since in general $T_1 \gg T_2$, the condition quoted for simplified calculation of T_1 and T_2 from the circuit elements is then also satisfied. The time constants T_1 and T_2 are linked with the characteristics of lightning impulse voltages by factors which depend upon the ratio T_s/T_r as follows:

$$T_s = k_2 T_2, \qquad T_r = k_1 T_1.$$

The values of these proportionality factors for the more important standard forms are:

T_s/T_r	1.2/5	1.2/50	1.2/200
k_1	1.44	0.73	0.70
k_2	1.49	2.96	3.15

For the voltage form 1.2/5 the requirement $T_1 \gg T_2$ is only partly fulfilled and hence, in this particular case, the approximate calculation often leads to considerable error. For the characteristics of switching impulse voltages we have:

$$T_{cr} = \frac{T_1 T_2}{T_1 - T_2} \ln T_1/T_2,$$

$$T_h \approx T_1 \ln \frac{2}{\eta} \quad \text{for} \quad T_h \geqslant 10\, T_{cr}.$$

If the conditions given above are only partly fulfilled, then the general solution for u(t) must be evaluated.

The shape of the voltage for lightning impulse voltages often deviates considerably from that calculated theoretically, particularly on the front and at the peak. This is caused by the inevitable inductance of the circuit elements and of the spatial setup, which can result in at least one point of inflexion in the front, and even superimposed oscillations. For a first look at the problem, an inductance L may be considered in series with R_d in the equivalent circuit, and the damping effect of the discharge resistance neglected ($R_e = \infty$).

To avoid interfering oscillations which make it difficult to give a value of $\hat{U}$, the circuit should be aperiodically damped. To this end, R_d may not be less than

$$2\sqrt{L\frac{C_s + C_b}{C_s C_b}} \; .$$

It is not easy to satisfy this condition in equipment for high voltages and high energy content.

1.3.4 Further Means of Generating Impulse Voltages

Short rectangular impulse voltages can be generated quite well with energy storage devices of the transmission line type. In a much used setup, a high-voltage cable is charged to a direct voltage U_0 via a high resistance and then discharged through a sphere gap onto an initially uncharged cable, at the end of which the test object is connected. The duration of the voltage impulse which develops across the test object is twice the travelling wave transit time of the charging cable; the peak value depends upon the impedance of the test object and is, at best, equal to U_0. In a different circuit high-voltage capacitors are switched onto a delay cable short-circuited at the end, the effective length of which can easily be varied to obtain impulse voltages of divers durations [*Winkelnkemper* 1965]. Voltage multiplication too can be temporarily realized using the transmission line type of energy storage device; the setup is in principle so arranged that the potential jumps caused by travelling waves on several lines add up at the test object. In an arrangement with two parallel line devices, suggested by *A. D. Blumlein* in 1941, voltage doubling is obtained. This Blumlein generator can be built as a double layer strip conductor, for example, the central electrode of which is charged at one end to U_0 with respect to the two outer ones. If one electrode pair at the beginning of the line is short-circuited, the resulting discharge wave causes a voltage jump of $2\,U_0$ at the test object connected between the outer electrodes at the end of the line. Generators of this kind have proved their merit particularly in plasma physics applications. An improvement of this method finally led to the development of "spiral generators" which can produce triangular voltage pulses of up to some 100 ns duration; their amplitude is a large multiple of the charging voltage [*Fitch, Howell* 1964].

To generate switching impulse voltages with times to crest in the millisecond range, besides the usual impulse generators, impulse-excited testing transformers may also be employed. An abrupt rise of the voltage in the excitation winding leads to a transient phenomenon between the transformer and the high-voltage side capacitors. The voltage produced at the latter is utilized as a switching impulse voltage. For impulse excitation supply is possible both from the a.c. mains [*Kind, Salge* 1965], as well as from charged capacitors [*Mosch* 1969]. Both methods have stood the test of practical application [*Anis* et al. 1975; *Thione* et al. 1975] and are illustrated by their basic circuit arrangement in Fig. 1.3-6a. Fig. 1.3-6b shows examples of possible high-voltage impulses.

The time dependence of the curve up to the peak value $\hat{U}$ takes the form $(1 - \cos \omega t)$, which can be described by the equivalent circuit of a series resonant circuit for this experimental setup. Here the transformer is replaced by its leakage inductance L_S, the high-voltage side capacitances are represented by C. Assuming a stable voltage source in

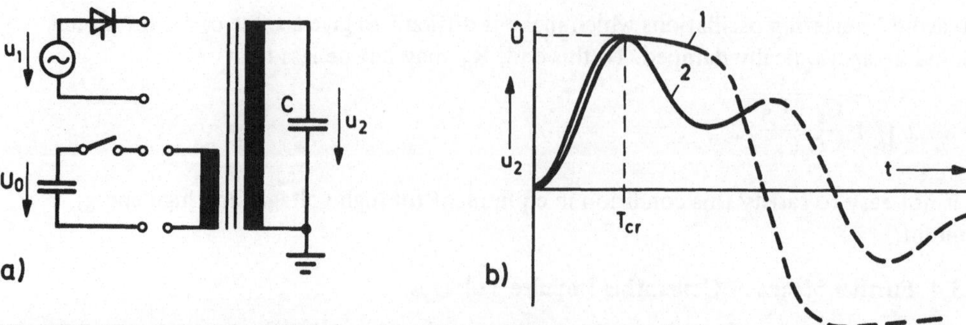

Fig. 1.3-6 Generation of switching impulse voltages using testing transformers
a) Basic circuit diagram for mains and capacitor excitation
b) Voltage curves for various switching devices
1 – thyristor 2 – mechanical switch

the excitation winding, for small damping of the circuit, the time to crest is estimated to be:

$$T_{cr} \cong \pi \sqrt{L_S C}.$$

The voltage curve after the peak remains more or less constant if a switching rectifier is used in the excitation circuit, e.g. a thyristor. A switch without the rectification, e.g. a mechanical switch, produces an oscillating output voltage. The dashed part of the curve shown in the figure is determined by the non-linear magnetization behaviour of the iron core.

This method is especially suitable for the generation of high switching impulse voltages with long times to crest. However, one should take care that transient phenomena within the testing transformer do not lead to overloading [*Wehinger* 1977].
Finally, it should be mentioned that high impulse voltages of short duration can also be generated using inductive circuits. For this purpose a high current is passed through the series combination of a high-voltage inductance and a switching device. The test object is connected in parallel with the switching device. If the resistance of the switching device increases strongly and the circuit current is maintained by the action of the inductance, a voltage pulse appears at the terminals of the test object. Exploding wires, for instance, have been found suitable as switching devices [*Salge* 1971].

Measurement of Impulse Voltages

1.3.5 Peak Value Measurement Using a Sphere Gap

The use of sphere gaps for the measurement of the peak value of high alternating voltages was described in section 1.1.6.From investigations on the breakdown of gases it is known that the development of a complete breakdown of such a system takes only a few μs at the most, if the applied voltage exceeds the peak value of the breakdown voltage $\hat{U}_d$ for alternating voltages. It follows that sphere gaps can be used to measure the peak value of impulse voltages, the duration of which is not too short. The limit is approximately $T_r \geqslant 50\,\mu s$.

It is assumed here that the air in the space between the spheres contains enough charge carriers to initiate the breakdown without delay after a definite field strength has been reached. By artifical irradiation, using UV sources or radioactive sources, the breakdown region can be sufficiently pre-ionised, so that the statistical scatter of the breakdown time is reduced. The relevant specifications therefore recommend that artificial irradiation be particularly used for the measurement of impulse voltages less than 50 kV.

A special feature of measuring the peak value of impulse voltages with sphere gaps is the fact that, on the basis of the occurrence or absence of a breakdown alone, one cannot ascertain how close the peak value $\hat{U}$ of the applied impulse voltage lies to $\hat{U}_d$. This can only be determined by repeated impulses.

To this end the amplitude of a sequence of impulse voltages is systematically varied until about half the impulses lead to breakdown, i.e., the breakdown probability $P(\hat{U})$ is about 50 %. For this impulse voltage we then have

$$U_{d-50} \approx \hat{U}_d \approx d\,\hat{U}_{d_0}$$

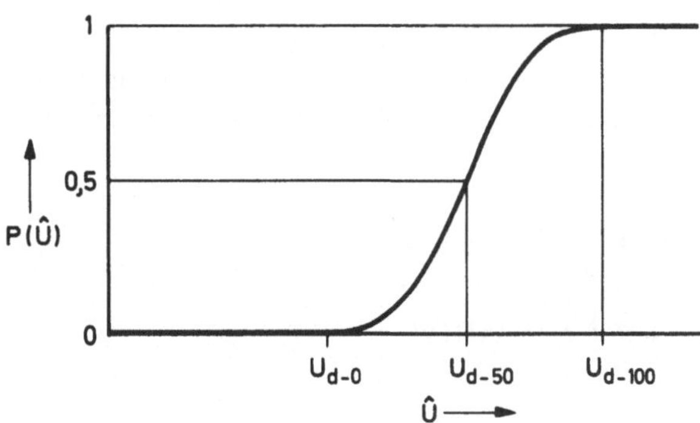

Fig. 1.3-7
Distribution function of the breakdown voltage of a sphere gap for impulse voltage

where d represents the relative air density and $\hat{U}_{d_0}$ is the breakdown voltage under standard conditions; the latter may be obtained from tables and depends upon the sphere diameter, polarity and spacing. The distribution function $P(\hat{U})$ of the breakdown voltage, shown in Fig. 1.3-7, may be determined by repeatedly stressing an electrode arrangement. It can be seen that the withstand voltage U_{d-0} and the assured breakdown voltage U_{d-100}, correspond ing to a breakdown probability of 0 % and 100 % respectively, can only be approximately defined and are therefore not suitable as characteristics.

Instead of the 50 % breakdown probability, which can usually only be accurately set for a large number of impulses, one can adjust to a value of $P(\hat{U})$ just below and another just above; the desired value U_{d-50} is then obtained approximately by interpolation, preferably graphically, where the ordinate is graduated as for a Normal distribution. The distribution function of the breakdown voltage is described in detail in Appendix A5, as well as a special method for the more exact determination of U_{d-50}.

1.3.6 Circuit and Transient Response of Impulse Voltage Dividers[1)]

The voltage shape of an impulse voltage is measured, using a cathode-ray oscilloscope (KO). The quantity to be measured is fed in via a coaxial measuring cable, the input end of which is connected to the secondary terminals of a voltage divider wired to the measuring point (test object). The divider leads, divider, measuring cable and the KO together constitute the measuring system. If the peak value $\hat{U}$ alone is to be measured, then a directly indicating electronic device may be connected instead of the KO.

a) Characteristics of the Transient Response

To investigate the response of measuring systems test functions are used. Characteristics are conventiently derived from the response to a step function. This method is suitable for theoretical as well as for experimental investigations.

One may consider the measuring system to be generally represented by a four-terminal network. A unit step voltage of amplitude $U_{1\infty}$ is applied as the input quantity:

$$u_1(t) = U_{1\infty} s(t).$$

The output voltage obtained is:

$$u_2(t) = U_{2\infty} w(t)$$

with $U_{2\infty}$ as rated value after the transient oscillations have died down. In these equations $w(t)$ is the step response to the unit step function $s(t)$. In linear systems the voltage $U_{2\infty}$ is proportional to $U_{1\infty}$. The expression $U_{1\infty}/U_{2\infty}$ is called the transformation ratio. An important characteristic for defining the response behaviour of a divider is the response time T, defined by the area:

$$T = \int_0^\infty [1 - w(t)]\, dt.$$

The simplest case of a four-terminal network with an aperiodic unit step response is shown in Fig. 1.3-8a. This kind of behaviour is named "RC behaviour".

Fig. 1.3-8b shows a four-terminal network with a unit step response incorporating a damped transient oscillation. When the response time is determined here, sub-areas of different sign result. The time T_1 can be considered a measure of the reproduction of the front of the step voltage, and the quotient T/T_1 a measure of the damping of the measuring system. The curve of the step response shown is described as "RLC behaviour".

In practical measuring systems much more complicated electrical circuits are very often in operation and quite different unit step responses can be encountered. As a result of large overshoot, the response time T can even become negative. For a wide band, and at the same time well-damped measuring system, T_1 shall be small and T/T_1 shall, if possible, tend to the value 1.

The origin of the unit step response can often only be poorly recognized in an oscillogram because of a slow initial oscillation or a superimposed oscillation. The time T_1 however

---

1) Comprehensive treatment in *Zaengl* 1970; *Hylten Cavallius* 1970; *Schwab* 1972; IEC Publication 60-4 (1977).

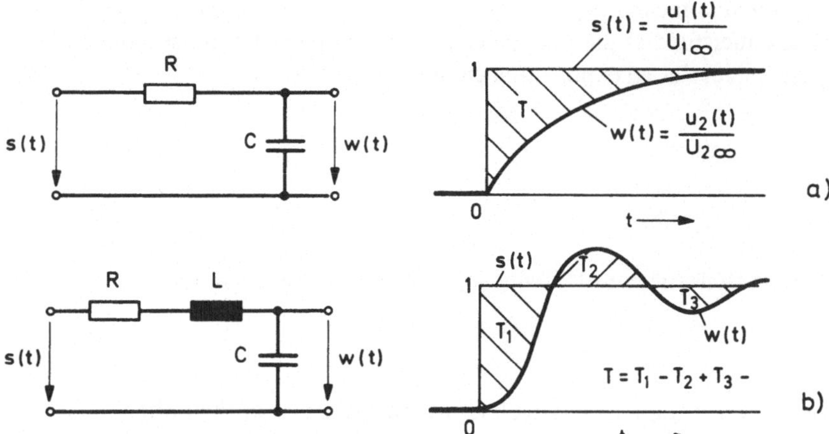

Fig. 1.3-8 Equivalent circuit and unit step response of voltage dividers
a) RC behaviour b) RLC behaviour

strongly depends upon the definition of the zero point. In these cases the origin of the unit step response is defined as the intersection of a linear elongation of the front with the zero line (IEC Publ. 60-4 (1977)).

If the transfer error of the voltage amplitude is not to exceed 5 %, then the following values apply as guiding values for the response time of the measuring system:

For full lightning impulses, and those chopped on the tail, of the form 1.2/50 $T \leqslant 200$ ns

For wedge-shaped impulse voltages rising nearly linearly up to T_d $T \leqslant 0.05\, T_d$

For theoretical investigations it is appropriate to apply the Laplace transformation, where the expressions given in Table 1.3-1 are valid for the parameters introduced:

Table 1.3-1

Description	Time domain	Laplace domain
Input voltage as step function	$u_1(t) = U_{1\infty}\, s(t)$	$U_1(p) = \dfrac{U_{1\infty}}{p}$
Output voltage as step response	$u_2(t) = U_{2\infty}\, w(t)$	$U_2(p) = \dfrac{U_{2\infty}}{p}\, G(p)$
Normalized transfer function		$G(p) = \dfrac{U_{1\infty}}{U_{2\infty}} \dfrac{U_2(p)}{U_1(p)}$
Response time	$T = \int_0^\infty [1 - w(t)]\, dt$	$T = \lim\limits_{p \to 0} \dfrac{1}{p}[1 - G(p)]$

In high-frequency impulse technology to characterize a step response the rise-time T_a is often used. This is understood as the time taken by the step response to rise from 10 % to 90 % of its peak value. For an exponential curve

$$w(t) = 1 - e^{-t/T}$$

the rise-time is:

$$T_a = 2.2\,T.$$

Under the same conditions the cut-off frequency of the system is given by

$$f_g = \frac{1}{2\pi T}\ .$$

During the measurement of rapidly varying high voltages, because of the finite velocity of propagation of electrical processes in the usually extended measuring systems, transit time phenomena can considerably affect the response behaviour. Voltage divider and lead must then be regarded as a whole [*Zaengl* 1970]. However, for the voltage variations encountered in lightning impulse and switching impulse voltages, the transient response of the measuring system is determined above all by the properties of the voltage divider. The following discussion shall therefore be restricted to the two most important basic types of impulse voltage divider, using simple equivalent circuits with lumped components to derive the response characteristics.

b) Resistive Voltage Divider

In measuring systems with resistive dividers, as in Fig. 1.3-9a, it is useful for the measuring cable K to be terminated at the KO with its surge impedance Z, thus loading the divider with an effective resistance of the same value. The most important disturbance of the ideal behaviour of the divider is brought about by the earth capacitance of the high-voltage branch R_1, which must necessarily be long for reasons of insulation at higher voltages.

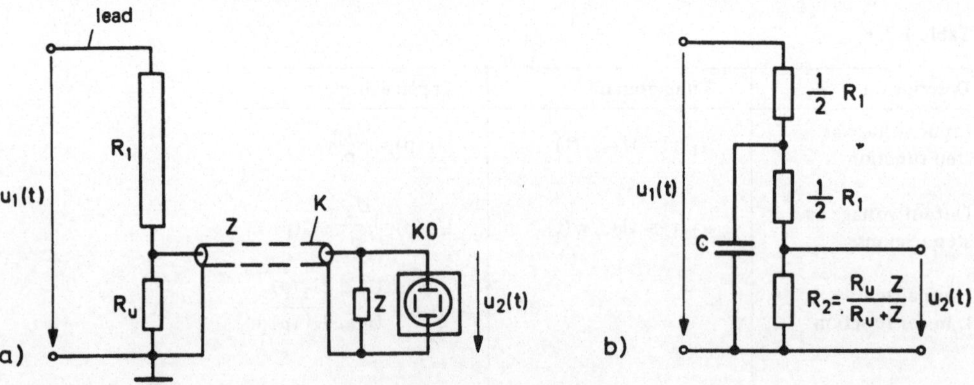

Fig. 1.3-9 Impulse voltage measuring system with resistive divider
a) Circuit diagram b) Equivalent circuit with earth capacitance

This earth capacitance is allowed for to a first approximation by the capacitance C in the equivalent circuit of Fig. 1.3-9b, which is connected at the centre of R_1.
Using the relationships introduced under a), the unit step response of this circuit can be derived as:

$$w(t) = 1 - e^{-t/T_R}.$$

For the time constant, using the approximation $R = R_1 + R_2 \gg R_2$, we have:

$$T_R \approx \frac{1}{4} RC.$$

The output voltage tends to the limiting value

$$U_{2\infty} = U_{1\infty} \frac{R_2}{R_1 + R_2}$$

w(t) corresponds to the curve shown in Fig. 1.3-8a, and the desired response time T is equal to the time constant T_R. Assuming a homogeneous distribution of the earth capacitances, it can be shown that C is equal to $\frac{2}{3}$ of the total earth capacitance C_e effective at R_1. It follows, therefore, approximately:

$$T \approx \frac{1}{6} RC_e.$$

For vertical cylindrical dividers a value of 15 ... 20 pF per metre height can be taken for C_e.
Hence, e.g., for a 1 MV divider with a resistance R = 10 kΩ and 3 m high, the earth capacitance is 60 pF and thus the response time T = 100 ns.
Resistive dividers are conveniently used for the measurement of steep impulse voltages of not too long a duration. Dividers for switching impulse voltages must be built with a large resistance R because of the heating and loading of the voltage source, which results in an unfavourable transient response for rapid voltage variations. For voltages above 1 MV, the practical construction of fast response resistive dividers becomes increasingly difficult, since one must try to compensate for the effect of the earth capacitances by increasing the coupling capacitances to the high-voltage electrode. One then has a capacitively controlled resistive divider which does however have a considerable capacitance parallel to the divider resistance. Moreover, this capacitance can be caused to oscillate with the inductance of the measuring circuit; in this way the system acquires RLC behaviour.

c) Capacitive Voltage Divider

In measuring systems with capacitive dividers, as in Fig. 1.3-10a, the measuring cable K cannot usually be terminated at the KO, since C_2 would discharge too rapidly because of the usual order of magnitude of the surge impedance ($Z \approx 100\ \Omega$). The series matching with Z indicated in the figure has the effect that only half the voltage at the divider tap enters the cable, yet this is doubled again at the open end, so that the full voltage will be measured at the KO once more. On the other hand the reflected wave may find matching at the cable input, since for very high frequencies C_2 acts as a short-circuit. The transformation ratio therefore changes from the value

$$\frac{C_1 + C_2}{C_1} \quad \text{for very high frequencies, to the value}$$

$$\frac{C_1 + C_2 + C_K}{C_1} \quad \text{for lower frequencies.}$$

However, the capacitance of the measuring cable C_K can usually be neglected compared with C_2.

Whilst the earth capacitance of capacitive dividers can be taken into account by a correction of the divider ratio, the response behaviour here is essentially determined by the inductance of the divider lead. As a first step, in the equivalent circuit of Fig. 1.3-10b an inductance L has been assumed in series with C_1. Using the relationships given in a), we obtain for this circuit:

$$w(t) = 1 - \cos\omega t, \quad \text{with} \quad \omega^2 = \frac{1}{L}\,\frac{C_1 + C_2}{C_1 C_2} .$$

For capacitive dividers an additional damping resistance is usually connected in the lead on the high-voltage side. The damping then present in the circuit causes the unit step response to assume the RLC behaviour, as shown in Fig. 1.3-8b for instance.

Particularly convenient behaviour results when damping resistances, the values of which approximately correspond to the aperiodic limiting case, are spread out and connected in series with the individual elements of the capacitive divider [*Zaengl* 1964]. This kind of damped capacitive divider acts for high frequencies as a resistive divider and for low frequencies as a capacitive divider. It can therefore be used over a wide range of frequencies, i.e. for impulse voltages of very different durations and also for alternating voltages. For

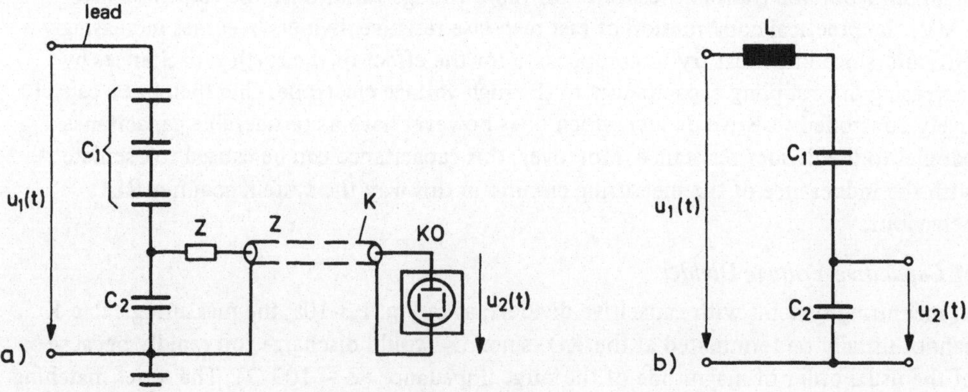

Fig. 1.3-10 Impulse voltage measuring system with capacitive divider
a) Circuit diagram b) Equivalent circuit with lead inductance

the calculation of these and similar divider circuits it is useful to think of the divider as a homogeneous iterative network and then solve the derived equations numerically.

1.3.7 Experimental Determination of the Response Characteristics of Impulse Voltage Measuring Circuits

For exact measurement of rapidly varying voltages the complete impulse voltage measuring circuit indicated schematically in Fig. 1.3-11 must be considered. Here the voltage $u_1(t)$ to be measured is the voltage at the terminals of the test object, whilst the measured value $u_2(t)$ corresponds to the curve on the KO screen. The response time of the entire measuring system, T_{res}, is obtained from the response time T of the divider with lead, the response time of the coaxial measuring cable T_K and the response time of the oscilloscope T_{KO}. If all three components show RC behaviour, to be verified in the case of the divider, then the resultant response time of the system can be calculated from the following equation:

$$T_{res} = T + T_K + T_{KO}.$$

The effect of the individual components on the resultant response time can be estimated from this relationship. T_K is usually much smaller than T, and can be neglected in the case of high-quality cables which are not too long. Impulse voltage oscilloscopes normally have a limiting frequency of over 50 MHz, from which $T_{KO} \leqslant 3$ ns can be calculated. In most cases the response behaviour of the divider determines the response time of the entire system, so that one may take $T_{res} \approx T$.

Investigations to determine the step response of a system can be carried out using high voltage as well as low voltage. In the first case one makes use of the voltage collapse of a spark gap with as homogeneous a field as possible, preferably operating at higher field strengths (compressed air, oil) to steepen the collapse. For measurements with a low-voltage square wave generator the signal, reduced by the divider in the divider ratio, requires a sensitive amplifying oscilloscope. One should therefore make sure that the latter's display behaviour does not differ appreciably from that of the KO used for high-voltage measurements, or that the difference is suitably taken into account.

Another method of determining the response time is by recording the impulse voltage-time curve of a known electrode configuration, using the measuring system to be investi-

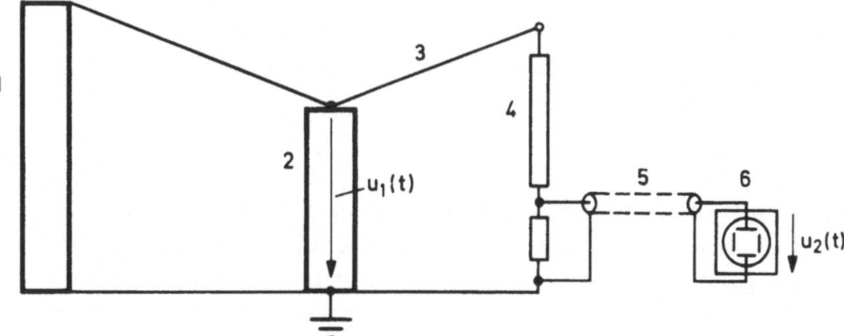

Fig. 1.3-11 Circuit of a complete impulse voltage measuring system. 1 Impulse voltage generator, 2 Test object, 3 Divider lead, 4 Divider, 5 Measuring cable, 6 KO

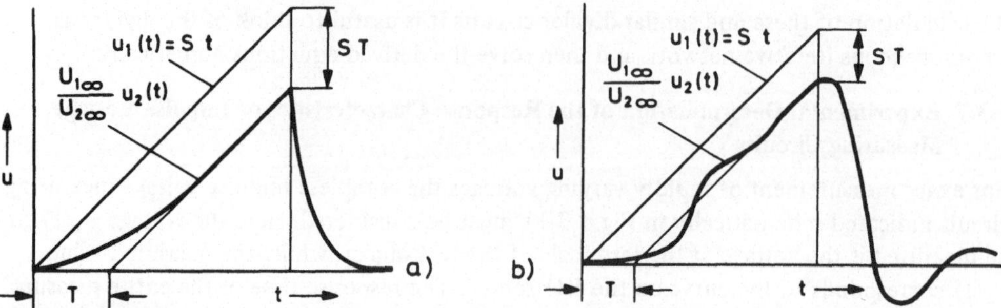

Fig. 1.3-12 Display of a wedge-shaped impulse voltage
a) System showing RC behaviour b) System showing RLC behaviour

gated[1]). The response time can be evaluated from a comparison of the measured impulse voltage-time curve, for linearly rising impulse voltages of constant rate of rise S, with the "true" impulse voltage-time curve of this same configuration. This method of response time determination is based upon the fact that the steepness of a wedge-shaped impulse voltage, after a certain transient and damping period, is reproduced accurately by all measuring systems in question. This behaviour is shown in Fig. 1.3-12. For a known transfer function G(p), for example, the curve can be calculated using the relationship

$$U_2(p) = \frac{U_{2\infty}}{U_{1\infty}} U_1(p)\, G(p).$$

For a divider corresponding to the equivalent circuit of Fig. 1.3-9b, with $u_1(t) = St$, we find:

$$u_2(t) = \frac{R_2}{R_1 + R_2} S\,[t - T(1 - e^{-t/T})].$$

If the steepness is accuarately reproduced, the response time can be determined from the voltage error ST. In practice several measurements are made with as many different values of S as possible, and the response times obtained are averaged. This method has the advantage that the measuring system can be investigated by a test method coming close to the actual requirements and at comparatively high voltages. Fundamental investigations of electrode configurations in air have shown that, for sphere gaps with only a weak inhomogeneous field, the true impulse voltage-time characteristic referred to standard conditions can be approximated by:

$$U_d = \hat{U}_{d_0} + \sqrt{2FS}.$$

Here $\hat{U}_{d_0}$ is the static breakdown voltage according to section 1.1.6 and F is the voltage-time area of the gap, as discussed in section 3.8.1b. Since the voltage-time areas of geometrically similar configurations are to a good approximation proportional to the static breakdown voltage, the above relationship allows easy conversion from known impulse

---

1) IEC Publ. 60-3, 1976: High-Voltage Test Techniques. High Voltage Measuring Devices

voltage-time characteristics to other configurations. As a test gap the IEC has suggested a single pole earthed sphere gap with D = 250 mm and s = 60 mm, and a static breakdown voltage $\hat{U}_{d_0}$ = 161 kV for standard conditions and negative polarity; the impulse voltage-time characteristic of this gap has been determined by international comparison measurements. These measured values are obtained from the above equation with sufficient accuracy, if the value 2 kV μs is substituted for F. For another sphere gap with the static breakdown voltage $\hat{U}^*_{d_0}$ the voltage-time area is given by:

$$F^* = F \frac{\hat{U}^*_{d_0}}{\hat{U}_{d_0}} .$$

1.4 Generation and Measurement of Impulse Currents

Rapidly varying transient currents of large amplitude as a rule appear in connection with high voltages, namely through the discharge of energy storing devices. They often develop as a consequence of breakdown discharge mechanisms and are frequently accompanied by large forces and high temperatures.

If these currents have a definite shape they are referred to as impulse currents; among other things, these are required for the simulation of lightning and short-circuit currents during tests on service equipment. Examples of the specific application of the physical effects of impulse currents are magnetic field coils for the confinement of plasmas, electrodynamic drives or gaps as impulse radiation sources.

The measurement of rapidly varying high currents is usually performed with measuring resistors, or with arrangements which exploit the inductive effect of the current to be measured.

1.4.1 Characteristic Parameters of Impulse Currents

Impulse currents can have very different shapes, depending upon their application and occurrence. Quite often impulse currents appear as aperiodic or damped oscillatory currents, and as alternating currents with a duration of only a few half-periods. The maximum instantaneous value of the current is denoted the peak value $\hat{I}$; characteristic parameters for the time dependence will be mentioned here only for impulse currents intended for testing.

To simulate currents produced by lightning strokes single, uni-directional impulse currents of short duration are used, which reach a peak value $\hat{I}$ rapidly without appreciable oscillation and then decrease to zero. The characteristics of these double exponential impulse currents are defined in accordance with the analysis parameters for impulse voltages given in Fig. 1.4-1a (IEC Publ. 60-2 (1973)). Usual values are $T_s = 4\ \mu s$, $T_r = 10\ \mu s$.

Rectangular impulse currents appear during the discharge of long transmission lines. The duration of these impulses is the time T_d during which the current stays larger than 0.9 $\hat{I}$ (Fig. 1.4-1b). Rectangular impulse currents with $T_d = 2000\ \mu s$ are often used for testing surge diverters.

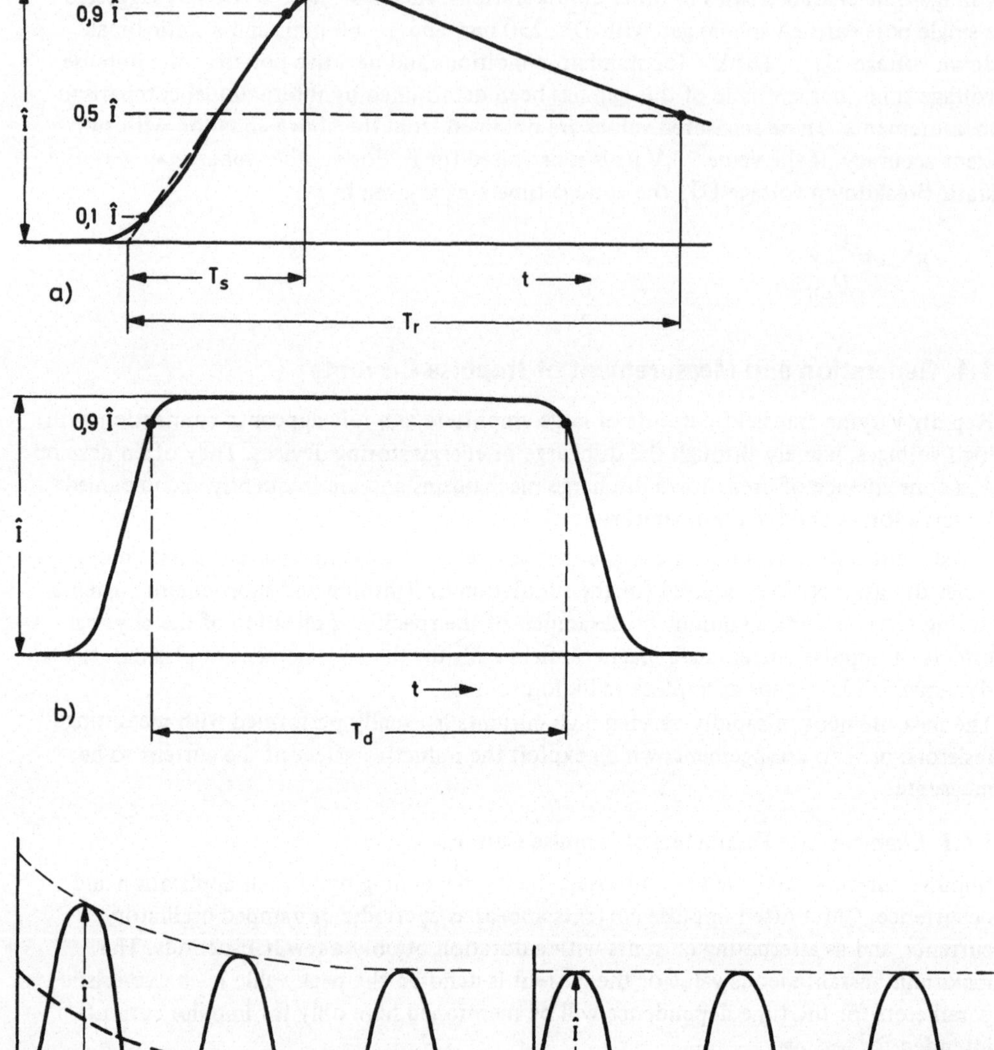

Fig. 1.4-1 Examples of impulse currents

a) Double exponential impulse current

b) Rectangular impulse current

c) Sinusoidal impulse current with exponentially decaying d.c. component

d) Sinusoidal impulse current without d.c. component

Impulse currents which occur in a.c. networks during short-circuiting are alternating currents which may be superimposed on an exponentially decaying d.c. component. Here the highest instantaneous value of the current determines the dynamic stress to which the components of the setup are subjected; it is named an impulse short-circuit current. For short-circuit tests on operating equipment a current shape as in Fig. 1.4-1c is aimed at, which represents particularly high stress. For an appropriate switching instant, a current can also appear without the d.c. component, as shown in Fig. 1.4-1d.

Generation of Impulse Currents [1)]

1.4.2 Energy Storage Systems

For the generation of high impulse currents the power which may be drawn from the power supply network is not normally sufficient to obtain a current of given shape and amplitude. In these cases one has to resort to energy storage systems which can be discharged with much greater power than is required to charge them. Capacitors, inductors, transmission line storages, rotating machines, accumulator batteries and even explosives are basically available as energy storage devices. The use of batteries and explosives is restricted to special cases only; these will not be discussed here further.

a) Capacitive Energy Storage

The energy stored in a capacitor of capacitance C at the voltage U_0 is given by:

$$W = \frac{1}{2} C U_0^2 .$$

It follows that the energy density in the dielectric stressed at the field strength E is given by:

$$W' = \frac{1}{2} \epsilon_0 \epsilon_r E^2 .$$

If one substitutes $\epsilon_r = 4$, $E = 1000$ kV/cm, the values achievable with oil-impregnated paper, one obtains $W' = 0.2$ Ws/cm^3. Capacitors are energy storage devices of high quality and extremely suitable for power amplification. They are able to store energy over a long period of time. The time constant for discharge through its own insulation resistance often reaches the order of hours. Consequently, a capacitive energy storage device can be charged by a source of low power.

The largest storage systems of this type were built for experimental investigations in plasma physics to generate high magnetic fields; their energy content is a few MWs and the charging voltage some tens of kV. When these systems discharge, currents of several tens of MA are obtained. Further fields of application are test setups for surge diverters, lightning current simulation and electro-hydraulic metal forming. Diverse installations and their working principles are described in the literature [*Mürtz* 1964; *Prinz* 1965; *Bertele, Mitterauer* 1970].

---

1) Comprehensive survey in *Craggs, Meek* 1954; *Sirotinski* 1956; *Früngel* 1965; *Knoepfel* 1970 and others

b) Inductive Energy Storage

The energy stored in a coil of inductance L for a current I_0 is given by:

$$W = \frac{1}{2} L I_0^2 .$$

For the energy density in the space filled with magnetic flux of density B, we have:

$$W' = \frac{1}{2} \frac{B^2}{\mu_0 \mu_r} .$$

Inductive energy storage systems are built with air-cored coils, since the maximum value of B is restricted to about 2 T in magnetic materials by saturation processes. The maximum value of the energy density is limited by the heating of the leads and by the magnetic forces. A value of B = 10 T can readily be achieved with coils under normal conductivity conditions, so that W' = 50 Ws/cm³. This value lies well above the energy density attainable in the electric field of a capacitor.
A major limitation in the application of inductive energy storage devices with conductors under normal conductivity conditions is the self-discharge time constant L/R, which is of the order of a few seconds and is very low compared with a capacitor system. Normally conducting storage coils therefore have to be charged with high power and can only be used as short-term storage. Long-term inductive storages may only be realized using superconducting coils.

c) Mechanical Energy Storage

In mechanical energy storage devices the energy is stored in moving masses and can be released by abrupt deceleration. The kinetic energy stored in a mass m moving with velocity v is:

$$W = \frac{1}{2} m v^2 .$$

In gyrating masses energy densities greater even than those of the magnetic field can be realized. The maximum value of the energy density is limited by the centrifugal forces. For a steel flywheel with a peripheral velocity of 150 m/s, a value of W' = 100 Ws/cm³ results.
As a rule, the rotors of suitably designed generators function as gyrating masses for mechanical energy storage devices; the conversion of their kinetic energy into electrical energy is achieved by deceleration. Generators of sinusoidal impulse currents are designed as synchronous generators and those for generating unipolar impulse currents usually as unipolar generators.
As a result of their high energy densities, mechanical energy storage systems can be built with capacities up to the order of 1000 MWs and owing to the very large self-discharge time constant, low-power charging equipment is quite adequate.

1.4.3 Discharge Circuits for the Generation of Impulse Currents

The aim of impulse current circuits is to generate a rapidly varying transient current of specified form and amplitude in a given arrangement. This may be needed to test the withstand capacity of operating equipment against stress by an impulse current, or to repeatedly trigger certain physical effects, such as the excitation of magnetizing coils. Analogous to impulse voltage circuits, the arrangement in which a given impulse current is to be produced shall be designated the test object.

a) Circuits with Capacitive Energy Storage

The equivalent circuit of an impulse current circuit with capacitive energy storage is shown in Fig. 1.4-2a. L_1 and R_1 represent the unavoidable inductance and ohmic resistance respectively of the impulse current circuit. If the test object P consists of a resistance R_2 and an inductance L_2 in series, and if one groups $L_1 + L_2 = L$ and $R_1 + R_2 = R$, then the current curves shown in Fig. 1.4-2b are obtained for different values of resistance R on ignition of the three-electrode gap F, most commonly used as a switch [*Deutsch* 1964; *Bertele, Mitterauer* 1970].

In this kind of discharge circuit the impulse current is critically influenced by the test object. The highest peak value of the current is reached for the case of low damping, i.e. when

$$R \ll 2\sqrt{\frac{L}{C}}.$$

In this case, with the capacitor charged to a voltage U_0, we have for the peak value of the discharge current:

$$\hat{i} \approx U_0\sqrt{\frac{C}{L}} = \sqrt{\frac{2W}{L}}.$$

The maximum rate of rise of the current at $t = 0$ is:

$$\left(\frac{di}{dt}\right)_{max} \approx \frac{U_0}{L}.$$

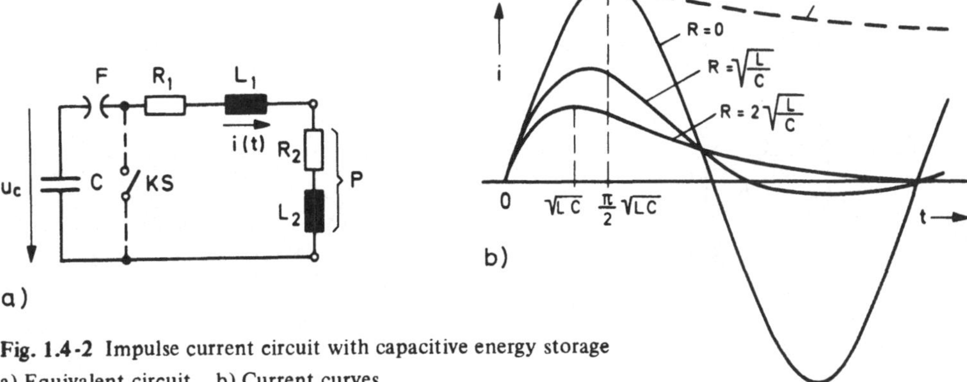

Fig. 1.4-2 Impulse current circuit with capacitive energy storage
a) Equivalent circuit b) Current curves

To increase the value of $\hat{I}$, and for maximum rate of rise of the current, one therefore tries to keep L low. This calls for compact assembly of the setup and, if necessary, parallel connection of several capacitor units. If the first current oscillation shall not take less than a prescribed duration, then besides decreasing L the value of C and hence W must be increased at the same time, so that the discharge frequency

$$f = \frac{1}{2\pi\sqrt{LC}}$$

does not become too high. In discharge circuits with low damping an aperiodic current curve can be achieved by the introduction of short-circuiters (KS in Fig. 1.4-2a) [*Bertele, Mitterauer* 1970].

b) Circuits with Inductive Energy Storage

Fig. 1.4-3 shows the equivalent circuit of an impulse current circuit with inductive energy storage and the most important current curves. The test object P here too consists of an inductance L_2 and a resistor R_2 in series. With the commutating switch SK closed, the storage coil L_s is charged to a current I_0 (from a source not shown in the figure) via the loss resistance R_s of the whole charging circuit. At $t = 0$ the switch SK is opened; the desired commutation of the switching current to the test object occurs when SK generates a voltage which is sufficiently high and tuned to the test object.

Neglecting the resistances R_s and R_2, from the requirement that the total magnetic flux must be the same before and after commutation, the condition

$$I_0 L_s = \hat{I}(L_s + L_2)$$

may be derived. Then for $L_s \gg L_2$ the full current amplitude can be expected in the load.

The realization of the switching device SK is a special technical problem of inductive energy storage circuits. In addition to arc switches, exploding wires or foils have proved good [*Salge* et al. 1970].

c) Circuits with Mechanical Energy Storage and Mains Supplied Setups

Impulse current arrangements with mechanical energy storage are mainly set up when very high energies are required for periods of up to one second. Mains supplied setups are also, in principle, mechanical energy storage systems because the energy is initially taken from the kinetic energy of the gyrating masses of the machines in the supply network.

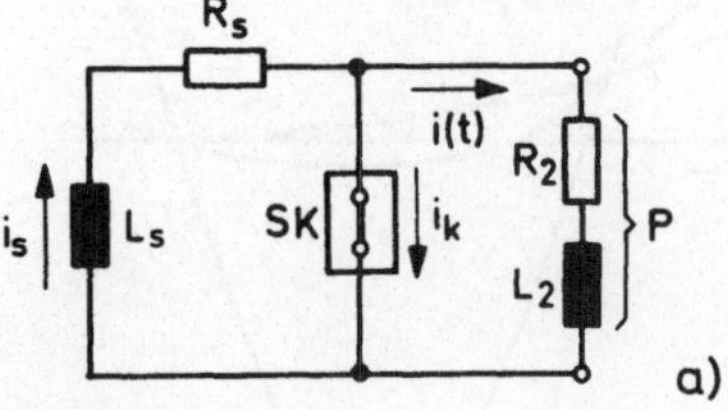

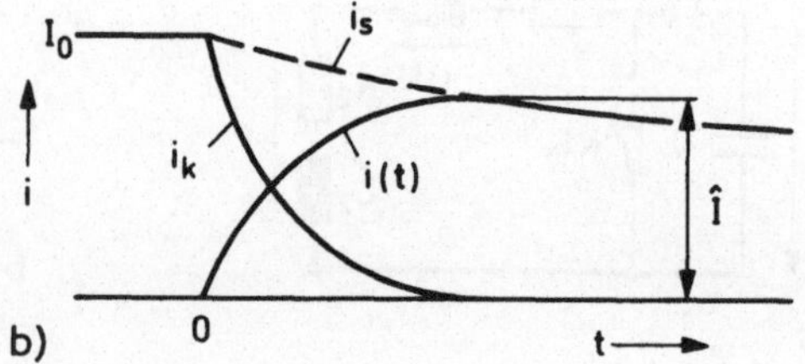

Fig. 1.4-3 Impulse current circuit with inductive energy storage
a) Equivalent circuit b) Current curves

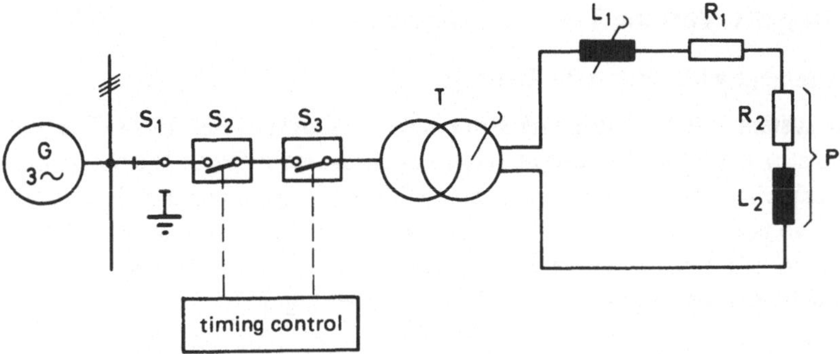

Fig. 1.4-4 Overall circuit diagram of a mains supplied impulse current setup

Particularly high powers and energies are required for testing high-voltage circuit breakers. Thus, very elaborate arrangements must be set up, often containing short-circuit generators to produce the impulse current [*Slamecka* 1966]. The requirements are usually much less stringent for basic investigations, for instance on arcs and contacts. In these cases, by connection to a three phase medium voltage network, powers of some MVA can be generated in comparatively less complicated setups, as will be shown in the example of Fig. 1.4-4.

An impulse current transformer T is connected to the network on the primary side via the earthing switch S_1, the series connection of the safety circuit breaker S_2 and the on-load circuit breaker S_3. The secondary side of the transformer is connected to the test object P via a coil L_1. Initially, after closing S_1, S_2 is closed. Shortly after S_3 is given its "on" command, S_2 can be given its "off" command again, so that both switches will only be in the "on" condition simultaneously for a short period. In this way it is possible to switch individual half-cycles of the mains voltage on using commercial circuit breakers. The short switching time has the great advantage that the supply network is only under load for short periods; very much more power can thus be drawn than would be possible for longer switching times.

d) Impulse Current Circuit with Transmission Line Type Energy Storage

Instead of concentrated capacitors and inductors a transmission line can also be used as energy storage for the generation of impulse currents. This is of some practical significance, particularly when nearly rectangular impulse currents are to be generated. The energy densities which can be achieved are the same as those given for capacitive energy storage systems. The current realized during the discharge, for the case of a short-circuit of the line, is determined by the surge impedance acting as the internal resistance of the system. In practical applications a multi-meshed iterative network is usually set up instead of a homogeneous transmission line; large impulse durations can then be more easily obtained.

Measurement of Rapidly Varying Transient Currents

1.4.4 Current Measurement with Measuring Resistors

A current measuring system with measuring resistor (shunt) is shown in Fig. 1.4-5. The voltage across the resistor R due to the current i(t) to be measured is fed to the oscilloscope KO via a measuring cable K. Termination of the cable with the surge impedance Z hardly affects the measuring voltage, if the condition $R \ll Z$ is satisfied. However, magnetic fields caused by the current to be measured and stray magnetic fields can induce voltages in the measuring circuit which are superimposed on the desired measuring signal iR. The following relationship is true:

$$u(t) = iR + L\frac{di}{dt} + \frac{d\Phi}{dt}.$$

Induced voltages due to stray magnetic fields Φ can usually be kept low by careful screening. An arrangement of the measuring circuit with low self-inductance L requires the current path within the screening to be such that minimum magnetic flux is linked by the loop formed at the measuring tap. With measuring shunts for high impulse currents of large duration, this is achieved only at considerable expense for thermal reasons.

Measuring shunts for impulse currents of short duration can be built with rise-times of a few ns order of magnitude. An example is shown in Fig. 3.12-5. The resistance element itself, depending upon the value of R, can be made of parallel carbon film resistors or low inductance wire resistors, of parallel resistance wires or resistance foils. Other modes of construction are described in the literature [*Sirotinski* 1956; *Gontscharenko* et al. 1966; *Schwab* 1972].

The potential link between the KO and the impulse current circuit, necessary when measuring shunts are used, is sometimes a disturbance in high-voltage measurements. It can often lead to the formation of earth loops and thus interfere in the measurements of very fast phenomena.

1.4.5 Current Measurement Using Induction Effects

If two circuits are magnetically coupled, as shown in Fig. 1.4-6a, the following expression is true:

$$u_2 = -i_2 R_2 - L_2 \frac{di_2}{dt} + M\frac{di_1}{dt}$$

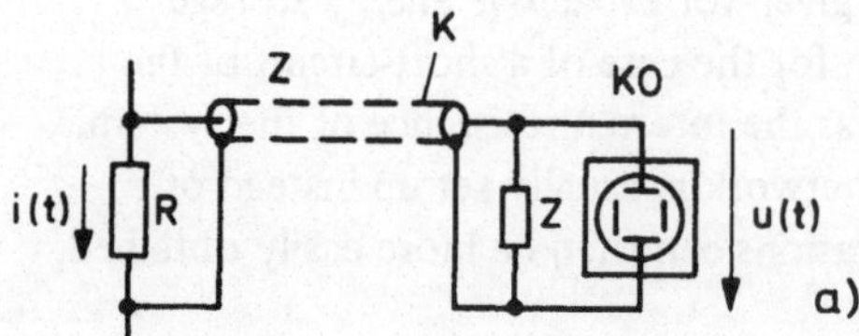

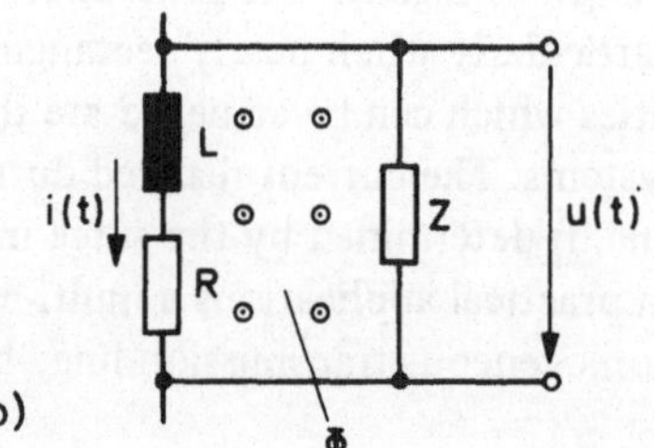

Fig. 1.4-5 Current measuring system with measuring resistor
a) Circuit diagram b) Equivalent circuit

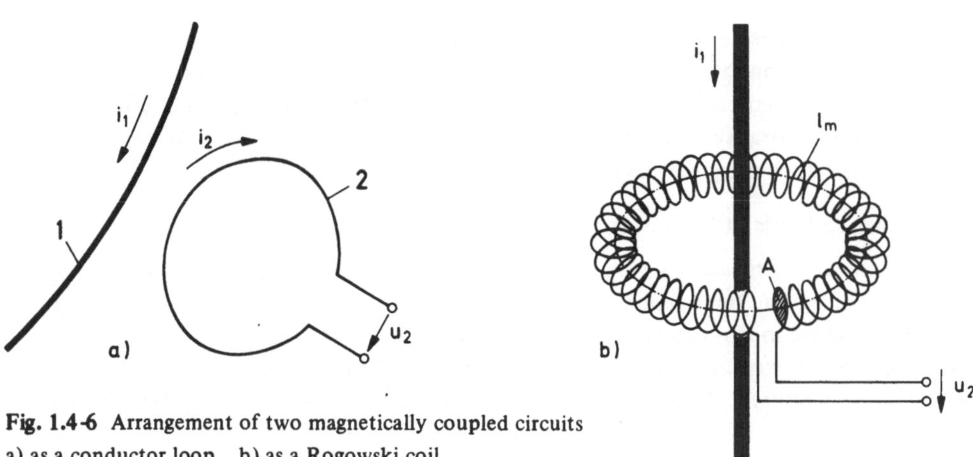

Fig. 1.4-6 Arrangement of two magnetically coupled circuits
a) as a conductor loop b) as a Rogowski coil

where R_2 and L_2 are the effective resistance and the self-inductance respectively, measurable at the terminals of circuit 2; M is the mutual inductance of circuits 1 and 2.

The first way of measuring i_1 with this arrangement refers the current measurement back to the measurement of u_2. A measuring system with a high internal resistance is connected to the terminals of circuit 2 and with $i_2 = 0$, we have:

$$u_2 = M \frac{di_1}{dt} .$$

In the other arrangement of the measuring loop, often in the form of the Rogowski coil, coil 2 encircles the conductor through which the current to be measured flows, in the manner shown in Fig. 1.4-6b. From the law of induction it follows directly that, for a uniformly wound coil with N turns, winding area A and length l_m

$$M = \frac{\mu_0 N A}{l_m} .$$

In order to obtain a parameter which is proportional to the current to be measured, the measuring voltage u_2 must be integrated. This can be done most easily by an RC circuit, but more elegantly by using an appropriately wired operational amplifier. To avoid measuring errors as a result of stray magnetic fields, the winding is usually of the criss-cross type.

The differential equation valid for Fig. 1.4-6a provides further means of current measurement if circuit 2, with the lowest possible effective resistance, is short-circuited. For $u_2 = 0$ and $R_2 = 0$ we have:

$$L_2 \frac{di_2}{dt} = M \frac{di_1}{dt} .$$

If the two circuits are magnetically closely coupled, $L_2 \approx M$ and

$$i_2 \approx i_1 .$$

This is a current transformer which is often built with a variable number of turns, i.e. with a transformation ratio differing from 1, and has an iron core for close magnetic coupling. Iron core current transformers are not suitable for measuring rapidly varying currents of short duration, but can be used with great accuracy for phenomena in the ms range and can be designed for large currents.

1.4.6 Other Methods of Measuring Rapidly Varying Transient Currents

As further means of measuring impulse currents one may mention those methods which make use of magnetic field dependent material properties. The Hall generator and magneto-optic elements are in this category [*Schwab* 1972]. For the Hall generator, through which a constant control current flows and which is permeated by the magnetic field of the current to be measured, the Hall voltage is directly proportional to the measuring current. This method became significant with the development of semiconductors with sufficiently large Hall constants. Magneto-optic methods of current measurement use the rotation of the plane of polarization in materials by the magnetic field which is proportional to the current (Faraday effect). The material is irradiated thereby with linearly polarized light. Both these methods are insensitive to overloading. Photo-luminescent diodes too, the momentary light emission of which is proportional to the current flowing through them, can be used for current measurement [*Wiesinger* 1969]. The advantage of these methods is that they allow galvanic isolation of the measuring setup from the main current circuit.

1.5 Non-Destructive High-Voltage Tests

When an insulation system is investigated, its breakdown voltage defines the upper limit of the voltage range. However, it is usually not possible to draw conclusions about the cause of breakdown discharge from a knowledge of the breakdown discharge voltage and the breakdown discharge tracks because, particularly in solid materials and for the application of powerful high-voltage sources, the insulation is destroyed in the region of breakdown. Dielectric tests which avoid a breakdown discharge are thus an important aid in the assessment of insulating materials and insulation systems.

1.5.1 Loss in a Dielectric

An ideal dielectric is perfectly loss-free and its behaviour in an electric field can be completely described by a real dielectric constant

$$\epsilon = \epsilon_0 \epsilon_r.$$

Conversely, loss always occurs in a real dielectric. The following may be considered to be the physical reasons for dielectric loss:

Conduction loss	P_l	by ionic or electronic conduction. The dielectric possesses finite conductivity κ.
Polarization loss	P_p	by orientation, boundary layer, or deformation polarization.
Ionisation loss	P_i	by partial discharges (PD) in or at an insulating arrangement.

These losses cause certain electrical effects, which can be utilized for non-destructive high-voltage tests. The most important measuring quantities are the following:

conduction current for direct voltage
dissipation factor for alternating voltage
partial discharge characteristics for alternating voltage.

For a given test object, these measuring quantities vary with the magnitude of the test voltage. They are also dependent in general upon test conditions, such as temperature and time, as well as upon the properties of the dielectric. Important material parameters are type, composition, structure, purity and previous history. These relationships furnish important information about a dielectric and in special cases may even indicate whether ageing of the dielectric has occurred.

When electrical equivalent circuits are set up for an insulation system, the object is to simulate complicated physical relationships by using circuits with similar electrical properties. However, many dielectric properties are non-linear, and yet one tries to get by with linear circuit elements in the equivalent circuit. These equivalent circuits should therefore be regarded with some caution, and be used only in accordance with their range of validity. Nevertheless, they are of no little use in many problems.

An attempt is made in Fig. 1.5-1 to present a general equivalent circuit for a dissipative dielectric. Whilst an ideal dielectric may be represented by a pure capacitance C_3, conduction losses can be taken into account by a resistor $R_0(\kappa)$ in parallel. Polarization losses produce a real component of the displacement current, which is simulated by the resistor R_3. Pulse-shaped partial discharges (PD) can be described by the right-hand branch. The gap spacing bridged by a gas discharge is represented by the capacitor C_1 with a gap F in parallel, whereby the repeated recharging of C_1 is effected either by a resistor R_2 or a capacitor C_2. Further details of this PD equivalent circuit will be given in section 1.5.4.

1.5.2 Measurement of the Conduction Current for Direct Voltage

In a steady field of strength $\vec{E}$ the current density $\vec{S}$, according to Ohm's law, for finite specific conductivity $\kappa = 1/\rho$ is:

$$\vec{S} = \kappa\,\vec{E}.$$

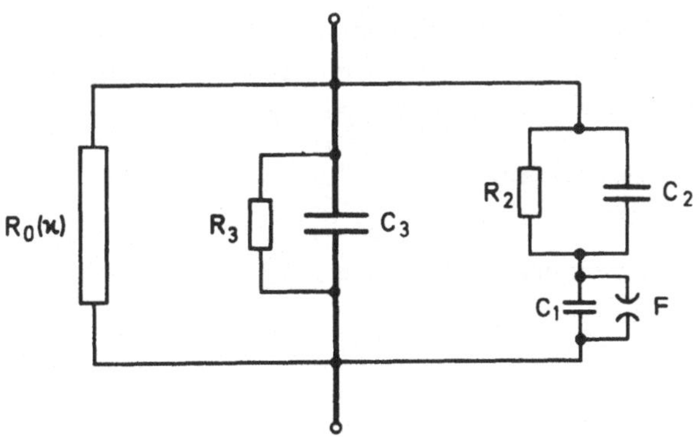

Fig. 1.5-1
Equivalent circuit of a dielectric with loss due to conduction, polarization and partial discharges

For the specific dielectric loss we then have:

$$P'_{diel} = \vec{E}\vec{S} = \kappa E^2 .$$

The conductivity of insulating systems comprising liquid and solid materials is mostly a result of ionic conduction and is therefore strongly dependent upon temperature, impurities and particularly upon the moisture content. The leakage resistance $R_0(\kappa)$ of an insulating system is determined by measuring the current when a constant direct voltage is applied. Since different types of conduction mechanism are working simultaneously, the measured result is also time-dependent; and so, for comparable values, the measurement should be performed at a definite time after switching the measuring voltage on, e.g. after 1 minute. For simple electrode geometry the specific quantities κ or ρ may be calculated from the value of the resistance.

A simple arrangement for investigating insulation plates is shown schematically in Fig. 1.5-2. The measuring voltage, usually 100 V or 1000 V, is applied between the electrode 1 and earth. The measuring electrode 2 is earthed via a sensitive ammeter; the guard-ring electrode 3, always necessary to eliminate boundary field effects and surface currents, is directly earthed. For most insulating materials κ lies in the region of $10^{-16} \ldots 10^{-10}$ S/cm, resulting in measuring currents of the order of picoamperes to nanoamperes. The measuring leads should be appropriately carefully screened.

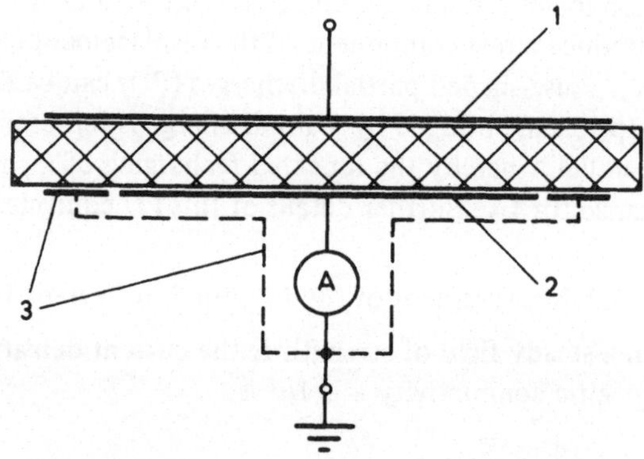

Fig. 1.5-2
Electrode arrangement to measure the conductivity of an insulation plate for direct voltage
1 High-voltage electrode
2 Measuring electrode
3 Guard-ring electrode and screening

Direct voltage measurements to determine the conduction current are not only important to derive the specific material properties but above all to check the condition of high capacitance insulating systems. This method has proved good, for instance, in the insulation control of large electrical machines during their period of operation.

1.5.3 Measurement of the Dissipation Factor for Alternating Voltage

a) Dielectric Loss and Equivalent Circuits [1)]

In an alternating field of strength $\vec{E}$ the current density $\vec{S}$ is generally given by:

$$\vec{S} = (\kappa + j\omega\tilde{\epsilon})\vec{E}.$$

1) Full description in *v. Hippel* 1958; *Lesch* 1959; *Andersson* 1964 and others

As a rule, polarization and ionisation losses also occur in a dielectric, besides the conduction loss. The dielectric constant $\tilde{\epsilon} = \epsilon_0 \tilde{\epsilon}_r$ is here no longer a real quantity. The dielectric dissipation factor $\tan\delta$ of an insulation is defined as the quotient of the real component I_w of the current and its imaginary, or reactive, component I_b:

$$\tan\delta = \frac{I_w}{I_b} = \frac{P_{diel}}{P_b} .$$

The dielectric losses comprise three parts, corresponding to the loss mechanism:

$$P_{diel} = P_l + P_p + P_i$$

for each of which an individual dissipation factor can be quoted, so that

$$\tan\delta = \tan\delta_l + \tan\delta_p + \tan\delta_i .$$

In a phasor diagram of the fundamental frequency components δ appears as the angle between the current flowing through the dielectric and its imaginary component; for small angles $\tan\delta$ is then equal to the power factor $\cos\varphi$.

If only conduction losses occur, then the defining equation gives the simple relationship:

$$\tan\delta_l = \frac{\kappa}{\omega\epsilon_0\epsilon_r} .$$

This part is inversely proportional to the frequency and can therefore be neglected at high frequencies. For supply frequency, on the other hand, each loss component can have considerable magnitude.

Instead of the dissipation factor, the expression

$$\tan\delta = \frac{\epsilon''}{\epsilon'}$$

is often introduced, where ϵ' is the real part of the relative permittivity and ϵ'' corresponds to the dielectric loss. Then by definition:

$$P_{diel} = P_b \tan\delta = \omega C \tan\delta\, U^2 .$$

If this equation is applied to a volume element, in the form of an infinitesimal cube, for the specific dielectric loss we then have:

$$P'_{diel} = \omega\epsilon_0\epsilon' \tan\delta\, E^2 .$$

When this equation is applied the approximation $\epsilon' \approx \epsilon_r$ can generally be used.

Fig. 1.5-3 shows two equivalent circuits for a dissipative dielectric at alternating voltage. Here the losses are simulated by resistances. The dissipation factor is given for the parallel circuit of Fig. 1.5-3a by:

$$\tan\delta = \frac{1}{R_p\,\omega C_p}$$

and for the series circuit of Fig. 1.5-3b by:

$$\tan\delta = R_s\,\omega C_s .$$

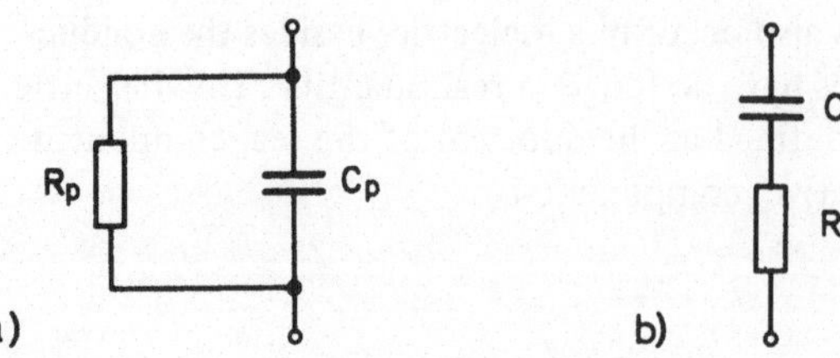

Fig. 1.5-3
Equivalent circuits for a dissipative dielectric for alternating voltage
a) Parallel equivalent circuit
b) Series equivalent circuit

For a fixed frequency both equivalent circuits are indeed equivalent and the elements can be converted accordingly. However, the frequency dependence is just the opposite in the two cases, and this illustrates the limited validity of these equivalent circuits.

With certain restrictions even the frequency dependence can be simulated by adding other elements to these circuits.

b) Measurement with the Schering Bridge

Measurement of the dielectric loss for alternating voltage is usually carried out in high-voltage technology using the bridge circuit devised by *H. Schering* in 1919 (Fig. 1.5-4).

The Schering bridge is an alternating current bridge made up of capacitors and resistors. The balancing elements to be adjusted are inside an earthed enclosure whilst the test object C_x and the standard capacitor C_2, which should be practically loss-free, are at high voltage. The null indicator N may only be sensitive to the fundamental frequency

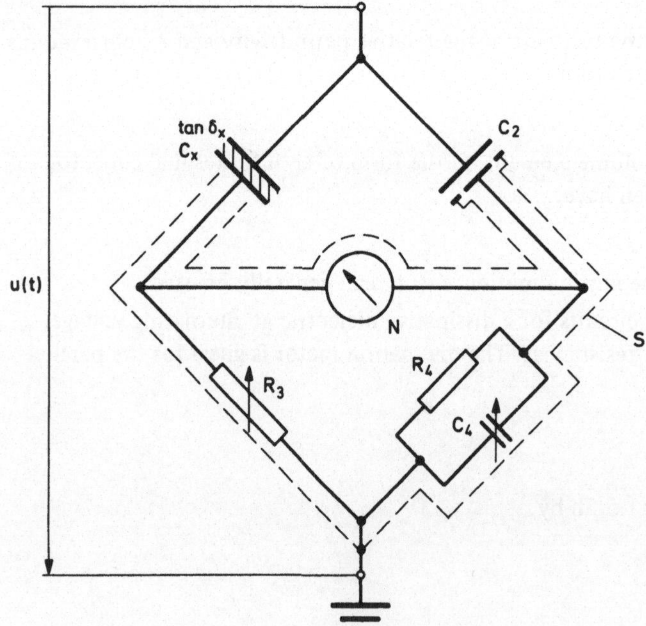

Fig. 1.5-4
Circuit of the Schering bridge
C_x Test object
C_2 Standard capacitor
R_3, C_4 Balancing elements
R_4 Fixed resistor
N Null indicator
S Screening

of the test voltage, which usually deviates from the sinusoidal. The corner points of the bridge must be provided with overvoltage protective devices so that, should the test object break down, overvoltages in the low-voltage circuit will be prevented.

The capacitance and dissipation factor of the test object are determined by adjusting the resistor R_3 and the capacitor C_4. For the balanced bridge, the following relation holds for the admittances of the bridge arms:

$$\tilde{Y}_4 \tilde{Y}_x = \tilde{Y}_2 \tilde{Y}_3 .$$

Taking the parallel equivalent circuit of Fig. 1.5-3a for the test object and substituting we have:

$$\left(\frac{1}{R_4} + j\omega C_4\right)\left(\frac{1}{R_x} + j\omega C_x\right) = \frac{j\omega C_2}{R_3} .$$

This equation must be satisfied by the real as well as the imaginary components, which corresponds to the Schering bridge balanced in amplitude and phase. The following relationships result:

$$\frac{1}{\omega C_x R_x} = \omega C_4 R_4$$

$$\frac{C_4}{R_x} + \frac{C_x}{R_4} = \frac{C_2}{R_3} .$$

For the required quantities we have:

$$\tan \delta_x = \omega C_4 R_4$$

$$C_x = C_2 \frac{R_4/R_3}{1 + \tan^2 \delta_x} \approx C_2 \frac{R_4}{R_3} .$$

Reflecting the significance of dissipation factor measurements at high voltages, the circuits and the equipment required to carry out these measurements have been the object of considerable further development compared with the basic circuit described above. One may mention here the additional circuitry to compensate the earth capacitances at the bridge corners, and current comparison circuits using transformers [*Potthoff, Widmann* 1965; *Schwab* 1972].

The null indicator should only record currents corresponding to the fundamental frequency of the alternating test voltage, since otherwise correct balancing is impossible; more sensitive electronic null indicators with oscillographic indication are preferred instead of mechanical vibration galvanometers.

The realization of the standard capacitor C_2 is a high-voltage technological peculiarity. For the derivation of the balanced condition relation it was assumed, a priori, that the dissipation factor of the standard capacitor is negligibly small compared with that of the test object. One therefore chooses capacitors with gas as an especially low loss dielectric. An arrangement with coaxial cylinder electrodes and compressed gas insulation, suggested in 1928 by *H. Schering* and *R. Vieweg*, is found to be particularly suitable for high voltages.

Exact knowledge of the capacitance C_2 of the standard capacitor is essential for measurement of C_x. C_2 should therefore be unaffected by external influences as far as possible.

1.5.4 Measurement of Partial Discharges at Alternating Voltages[1])

When the breakdown discharge strength of an insulation system is locally exceeded, either full or partial breakdown occurs, depending upon whether a discharge channel of low resistance develops between the electrodes or not. In the case of a partial breakdown the conditions for a stationary discharge are often not fulfilled as a consequence of insufficient energy input, and so only a short discharge pulse results. If the dielectric possesses self-healing properties in the overstressed region, and if the electric field builds up again, pulse-shaped partial discharges occur. Partial discharges may however also be pulse-free when they occur in conjunction with a stationary gas discharge.

Pulse-shaped and pulse-free partial discharges can occur for all types of voltage. However, with regard to the practical significance, only partial discharges at alternating voltages shall be discussed below, although the treatment of external partial discharges in particular is also valid for direct voltages.

a) External Partial Discharges

Collision ionisation sets in in gases when the onset voltage is exceeded at electrodes of small radius of curvature. In a strongly inhomogeneous field electron avalanches and photo-ionisation produce imperfect breakdown channels which, for alternating voltage, must re-ignite after each extinction of the partial discharges at the voltage crossover. This phenomenon, denoted external partial discharge or corona discharge, has great practical significance above all for overhead high-voltage transmission lines, because the maintenance of the discharge uses up energy (corona losses) and the current pulses produced generate electromagnetic waves (radio interference). External partial discharges also occur in test circuits for non-destructive high-voltage testing, and impede the detection of internal partial discharges which could possibly endanger the dielectric. External partial discharges can also occur in liquid insulators or at the interfaces of solid insulating materials; in the long run they will cause weakening of the insulation there, which might subsequently lead to a complete breakdown.

Fig. 1.5-5 shows a point-plane electrode configuration as a typical example of a setup with external partial discharges, as well as a grossly simplified equivalent circuit for pulse-shaped partial discharges. C_1 represents the capacitance assigned to the respective gas space breaking down and will be completely discharged when the ignition voltage U_z of the gap F is reached. The charge carriers formed at the point wander into the field, causing a certain conductivity which is represented in the equivalent circuit by R_2. Finally, C_3 is a parallel capacitance given by the electrode arrangement.

Assuming that $R_2 \gg 1/\omega C_1$, the current through R_2 is:

$$i_2 = \frac{u(t)}{R_2} .$$

[1]) Comprehensive treatment in *Kreuger* 1964; *Stamm, Porzel* 1969; *Schwab* 1972 and others

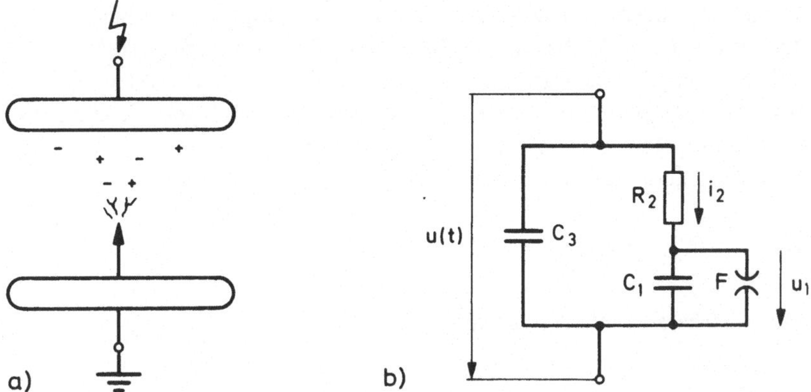

Fig. 1.5-5 Arrangement with external partial discharges and equivalent circuit
a) Point-plane electrode configuration
b) Equivalent circuit

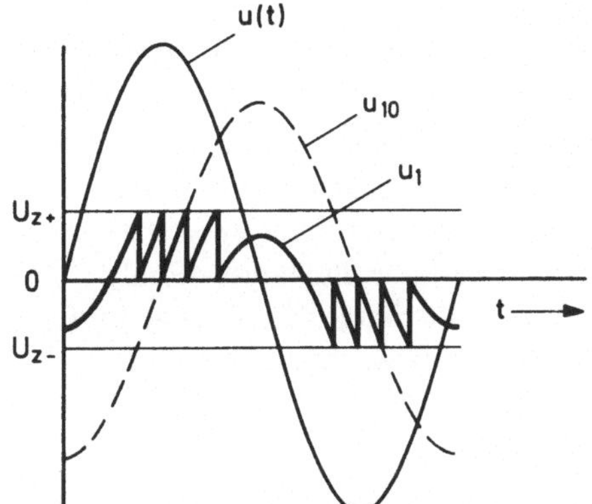

Fig. 1.5-6 Voltage curves in the equivalent circuit for pulse-shaped external partial discharges

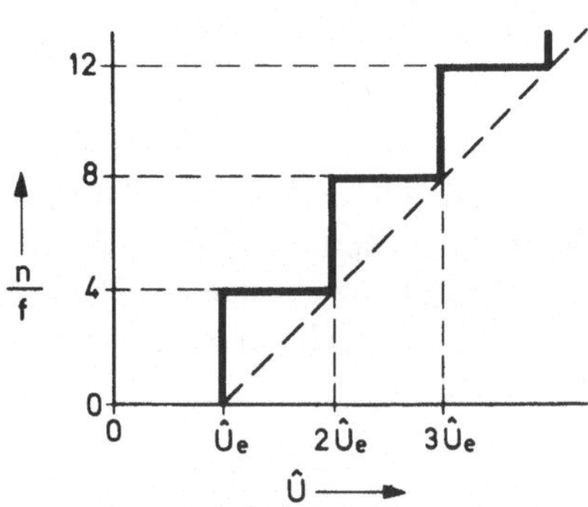

Fig. 1.5-7 Pulse repetition rate of partial discharges

If, as drawn in Fig. 1.5-6,

$$u(t) = \hat{U} \sin \omega t$$

is written for the test voltage, the open circuit voltage at C_1 at the end of the transient phase is:

$$u_{10} = \frac{\hat{U}}{\omega C_1 R_2} \sin(\omega t - \pi/2).$$

If the peak value of the test voltage just reaches the onset voltage

$$\hat{U}_e = \omega C_1 R_2 U_z$$

then the ignition voltage U_z appears across F and C_1 will be discharged in one burst. For increasing test voltage u (t), C_1 will subsequently be recharged by a voltage parallel in shape to u_{10} until U_z is reached once more, and so on. From the curves drawn for the voltage u_1 one can see that the partial discharge impulses occur predominantly in the peak of the test voltage. The pulse repetition rate n as a function of $\hat{U}$ is shown in Fig. 1.5-7. The dashed straight line is a good approximation for high pulse repetition rates:

$$n \approx 4f \frac{\hat{U} - \hat{U}_e}{\hat{U}_e} .$$

During each individual discharge a quantity of charge

$$Q_1 = C_1 U_z = \frac{\hat{U}_e}{\omega R_2}$$

is compensated for in F. This charge is then fed to C_1 again via R_2 by the voltage source and can be recorded as:

$$\Delta Q = Q_1 .$$

The prefix Δ is intended to show that this is a quantity corresponding to a complete pulse, yet with duration very short by comparison with the test voltage period.

The given equivalent circuit, by adding extra elements, can be made to correspond better to the physical behaviour of a real setup with external partial discharges. For example, one should note that the value of the onset voltage depends in many cases upon the polarity. The addition of a rectifier parallel to C_1 enables periodic discharges of only one polarity to be considered. The rectifier provides for the discharge of the capacitor C_1 during the half-cycles in which no partial discharges occur, because the onset voltage is too high.

b) Internal Partial Discharges

Where cavities are present in the liquid or solid dielectric of insulation systems, the field strength is greater inside than in the surrounding medium. When the voltage across the cavity exceeds the ignition voltage a partial breakdown results. Particularly for alternating voltages of sufficient amplitude a normally pulse-shaped discharge occurs in the cavity. The surrounding dielectric may deteriorate due to the long-term effect of these internal partial discharges, and under certain circumstances it can even be destroyed by total breakdown as a result of an erosion mechanism.

As a typical example of an electrode arrangement with internal partial discharges Fig. 1.5-8 shows an insulation system with a solid dielectric containing a gaseous cavity. The figure also shows the equivalent circuit for pulse-shaped partial discharges, suggested in 1932 by *A. Gemant* and *W. v. Phillippoff.* C_1 corresponds to the cavity capacitance which discharges completely via F when the ignition voltage U_z is reached. C_2 corresponds to the capacitance in series with the cavity and C_3 is the parallel capacitance of the arrangement. For a sinusoidal test voltage the open circuit voltage across C_1 is given by

$$u_{10} = \frac{C_2}{C_1 + C_2} u(t) = \frac{C_2}{C_1 + C_2} \hat{U} \sin \omega t.$$

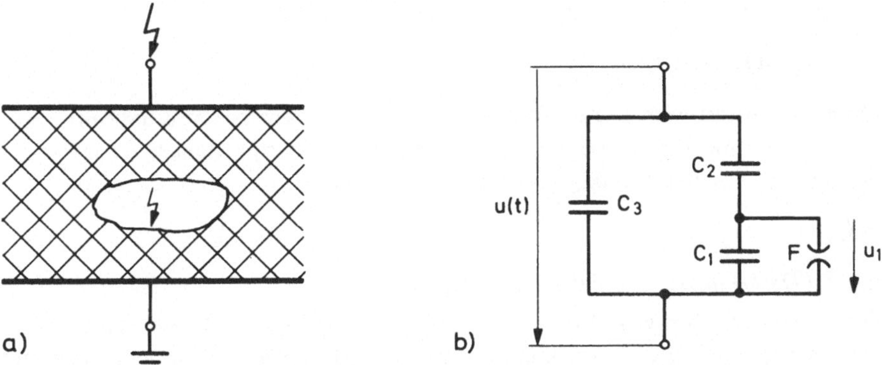

Fig. 1.5-8 Arrangement with internal partial discharges and equivalent circuit
a) Test object with cavity b) Equivalent circuit

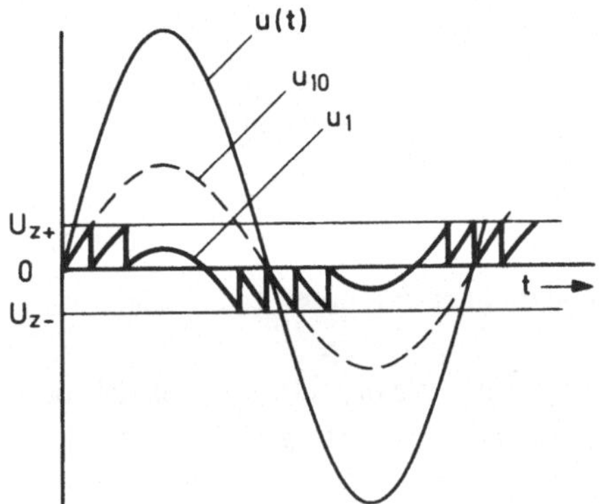

Fig. 1.5-9
Voltage curves in the equivalent circuit for pulse-shaped internal partial discharges

The peak value of the test voltage reaches the onset voltage $\hat{U}_e$ when the peak value of the open circuit voltage is just equal to U_z. It then follows that:

$$\hat{U}_e = \frac{C_1 + C_2}{C_2} U_z.$$

If the test voltage is greater than the onset voltage, repeated charging of C_1 occurs, as shown in Fig. 1.5-9. It is seen that the partial discharge pulses occur predominantly in the region of the test voltage crossover.

Here too the same relationship holds for pulse repetition rate as already given in a) and shown graphically in Fig. 1.5-7. The different phase relation of external and internal partial discharges is an important distinguishing characteristic of these two phenomena.

The charge compensated at the discharge site for each discharge is:

$$Q_1 = (C_1 + C_2) U_z,$$

whereas only the "apparent charge" Q_{1s} is fed to C_2:

$$Q_{1s} = C_2 U_z \neq Q_1 .$$

Here therefore, in contrast with a setup with external partial discharges, it is basically impossible to measure the actual charge Q_1 for unknown partial capacitances. The measurable charge Q_{1s} is denoted ΔQ in analogy with 1.5.4a:

$$\Delta Q = Q_{1s}.$$

c) Experimental Determination of Partial Discharges

An equivalent circuit which is valid for external and internal partial discharges is incorporated in the partial discharge test circuit reproduced in Fig. 1.5-10. A generator supplies a current pulse with impressed charge ΔQ for the short time $\Delta t \to 0$; this acts initially only on the test object capacitance C because of the impedances in the connecting leads. With the source current i_Q we have:

$$\Delta Q = \int_{\Delta t \to 0} i_Q \, dt = \text{const.}$$

Due to this the voltage across the test object suddenly changes by an amount:

$$\Delta U = \frac{\Delta Q}{C} .$$

This represents a very rapid process inside the test object, since the presented model contains no elements which could delay recharging.

For the subsequent transient processes in the test circuit only the capacitor C_k of the high-voltage circuit and the measuring resistor R are of significance. Referring to Fig. 1.5-10, one may now consider which of the partial discharge quantities can be measured at R. For the transient processes the nodal equation is:

$$i_{PD} = i_C + i_Q$$

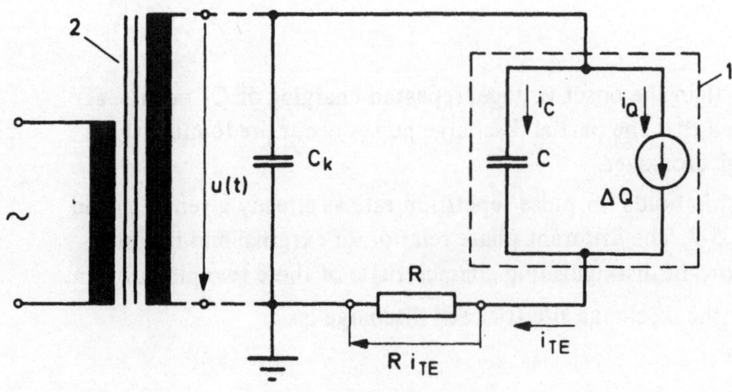

Fig. 1.5-10
Test circuit for measuring partial discharges
1 Test object with external or internal partial discharge
2 Test transformer
R Measuring resistor
C_k Coupling capacitor

and the loop equation is:

$$i_{PD} R + \frac{1}{C_k} \int_0^t i_{PD}\, dt + \frac{1}{C} \int_0^t i_C\, dt = 0.$$

By eliminating i_C it follows for the current pulse with impressed charge supplied by the generator:

$$\Delta Q = \left(1 + \frac{C}{C_k}\right) \int_0^t i_{PD}\, dt + i_{PD} RC.$$

The condition ΔQ = constant must be satisfied at all times t; this requirement is fulfilled by the Dirac pulse assumed for i_Q. For $t \to 0$ it follows from the above equation that i_{PD} jumps to a finite value at the instant of the pulse. Since there is no contribution from the integral at this instant, the initial value of i_{PD} is obtained from:

$$\Delta Q = i_{PD} RC = \Delta U_{PD} C.$$

A voltage jump ΔU_{PD} thus appears at the measuring resistor R. For $t \to \infty$, i.e. for times after decay of the transient process, the product $i_{PD} RC$ vanishes and we have:

$$\Delta Q = \left(1 + \frac{C}{C_k}\right) \int_0^{t \to \infty} i_{PD}\, dt = \left(1 + \frac{C}{C_k}\right) \Delta Q_{PD}.$$

Here ΔQ_{PD} represents the pulse-shaped charge flowing through R. Most of the measuring methods evaluate quantities which can be determined at the measuring resistor R when pulse-shaped partial discharges appear. When C is known, ΔQ may be determined from ΔU_{PD}; with the additional knowledge of C_k, ΔQ may also be determined from ΔQ_{PD}. The expression $(1 + C/C_k)$ corresponds to a transformer ratio of significance to the measuring sensitivity.

When partial discharge measurements are performed, a special coupling capacitor may be taken for C_k and the measuring resistor R can then be connected in the earth lead of C_k. However, the high-voltage side capacitances of the system, especially the winding capacitances of the transformer, are usually of sufficient magnitude, so that one can do without a special coupling capacitor.

For an adequate value of the coupling capacitance ($C_k \gg C$) the measured charge ΔQ_{PD} becomes equal to the charge ΔQ which is fed to that part of the insulation containing the partial discharge site. The transformer ratio may be experimentally determined in the setup with the aid of a pulse generator supplying impressed charge pulses [*Mole* 1970].

Since with ΔU_{PD} and ΔQ_{PD} we are in fact dealing with statistically varying processes, it is usual to determine the mean value of the measured quantities in these measurements (IEC Publ. 270, 1968). Under favourable conditions, knowing the trends of certain partial discharge parameters such as charge, pulse repetition rate, energy and time-dependence of the partial discharge pulses during stressing, the effect of internal ionisation processes on the quality of an insulation system can be assessed. On the other hand, one may not expect to make a reliable prediction of the life expectancy of an insulation system based upon a single measurement only [*Leu* 1966; *Kind, König* 1967; *Kodoll* 1974].

Measurement of the dissipation factor yields evidence about the losses occuring in the total electrically stressed volume; here, apart from the basic losses $P_l + P_p$, the partial discharge losses P_i too are determined. The considerations above using the equivalent circuit apply to pure pulse-shaped partial discharges. With the dissipation factor measurement, unlike many other methods, one also obtains the loss contributions caused by partial discharges which are not pulse-shaped. Nevertheless, the dissipation factor measurement is usually not sufficiently sensitive to locate single partial discharge sites in an insulation system.

2 Layout and Operation of High-Voltage Laboratories

2.1 Dimensions and Technical Equipment of the Laboratories

The dimensions and equipment of a high-voltage laboratory[1] are primarily determined by the magnitude of the voltage to be generated. A second important feature is the intended application e.g. for teaching purposes, as a testing or research laboratory.

2.1.1 Stands for High-Voltage Practicals

Practicals are laboratory exercises which give the students an opportunity to conduct set experiments under supervision. The experiments would generally be performed in small groups of three up to a maximum of six participants. The experimental stands described below are designed for this kind of practical.

In order to accommodate a large number of students, more stands must be available in which experiments can be conducted simultaneously. A useful guide for the setup of a practical laboratory would be approximately 1 experimental installation for every 20 students. The number of stands so derived imposes a certain restriction upon the voltage amplitude for economic reasons, which is also expedient with regard to clearer arrangement, and with that safety, of smaller setups.

If the maximum alternating voltage is restricted to 100 kV and the power ratings to between 5 and 10 kVA, the experimental stands could be set up in rooms with a normal height of 2.5 m. Moreover, the weight of the required construction elements, with the exception of the testing transformer, would be low enough to allow transport without crane facilities. Since most of the basic physical phenomena can already be observed within a voltage range of about 100 kV a.c., the restriction to this value does not impose any appreciable limit on the choice of experiments to be carried out. If necessary, the scope of the practicals could be widened by some demonstration experiments at very high voltage.

As an approved example, one of the five identically set up experimental stands for high-voltage practicals at the High-Voltage Institute of the Technical University Braunschweig will be described on the basis of Fig. 2.1-1. The protective barrier 1, consisting of wire mesh fixed to a metallic frame, is provided with a lockable door 2, near which work table 3 and control desk 4 are arranged. Inside the barrier is a working platform of two welded steel frames 5 with a covering of four hardwood panels. The steel frames serve as earthing points.

The high-voltage circuits are set up on the working platform; construction elements and accessories which are not required may be stored in trolley-drawers underneath. Further details may be taken from Fig. 2.1-2. For example, flexible cables for control and measuring purposes are already laid between the control desk and the working platform and need only be connected to the construction elements.

---

1) Comprehensive treatment in *Marx* 1952; *Craggs, Meek* 1954; *Prinz* 1965; *Leroy, Gallet* 1975 and others

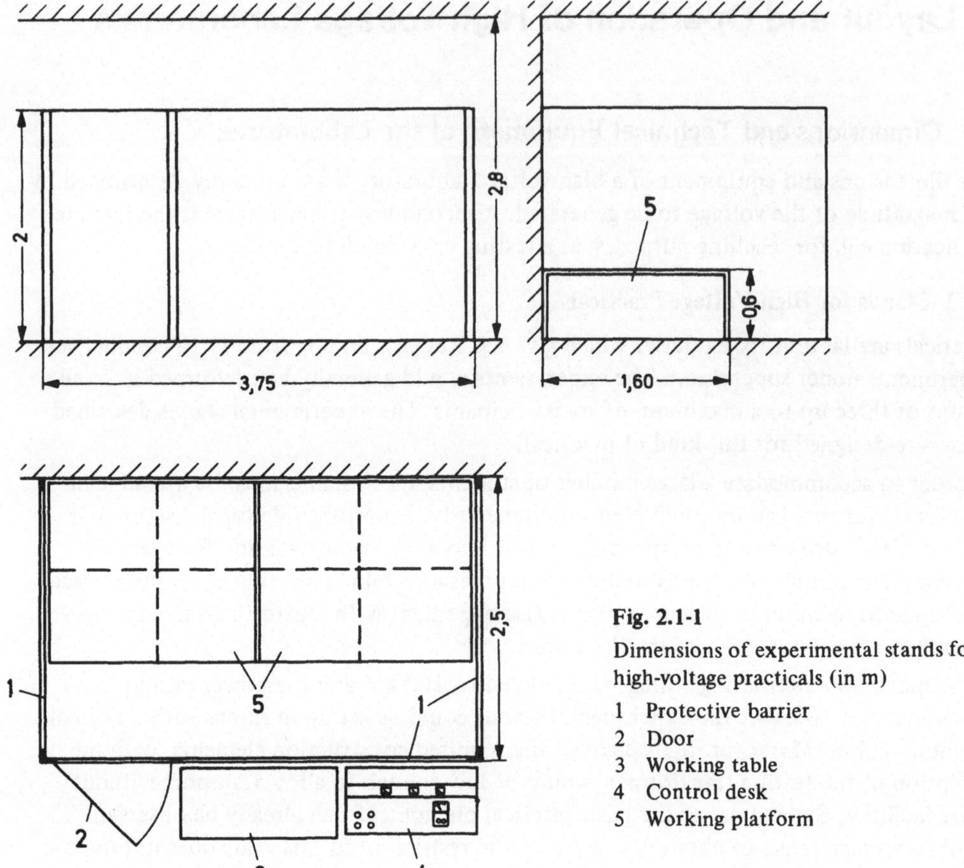

Fig. 2.1-1
Dimensions of experimental stands for high-voltage practicals (in m)
1 Protective barrier
2 Door
3 Working table
4 Control desk
5 Working platform

2.1.2 High-Voltage Testing Bays

In planning these, it should be considered that they are often responsible for an appreciable portion of the total investment and personnel costs. The introduction of partially automated measuring and protocolling devices permits considerable saving of costs.

Testing bays where routine and type-tests on manufactured high-voltage equipment are to be carried out, are usually adapted to a particular kind of test. They should constitute an integral part of the production line.

In practice, the operating voltage of the equipment to be tested influences the spatial arrangement of the testing bay. This is because, with respect to the minimum clearances to be maintained in a test setup (see Table 2.1-1), the clearance of the test room must, for high voltages, be considerably greater than the height of the production rooms. In consequence, a test object with an operating voltage over 220 kV would, for constructional reasons alone, require its own testing hall. In this case electromagnetic screening and black-out facilities for the room could also be easily introduced. Fig. 2.1-3 shows, as an example,

Fig. 2.1-2 Experimental stand for a high-voltage practical (Dimensions according to Fig. 2.1-1, Photo by *E. Sitte*, Braunschweig)

the possible layout of the testing hall for a factory in which transformers up to 400 kV are produced.

Table 2.1-1 Guiding values for test voltages and minimum clearances for test setups

Working voltage kV	Alternating voltage kV	Lightning Impulse voltage kV	Switching Impulse voltage kV	Minimum clearance m
30	85	170	–	–
110	260	550	–	–
220	505	1050	–	–
400	640	1425	900	4
765	960	2300	1300	12
1100	1400	2800	1800	20
1500	1900	3500	2200	30

Numerous technical disadvantages prevail when carrying out high-voltage tests in open air. In addition, there is uncertainty in planning the test schedule due to weather conditions, which is generally intolerable, particularly in factory testing bays. As a rule therefore, one prefers an indoor setup.

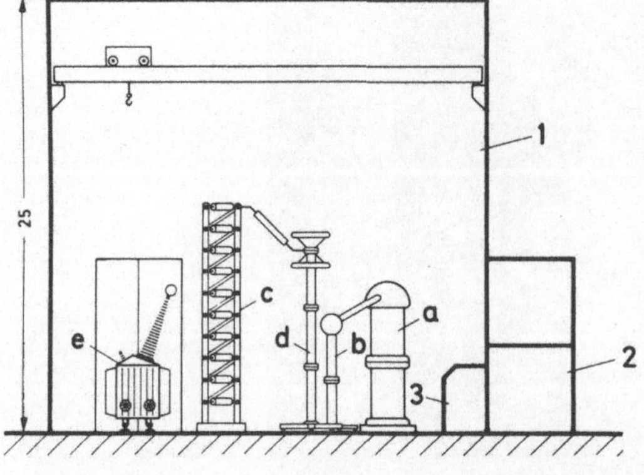

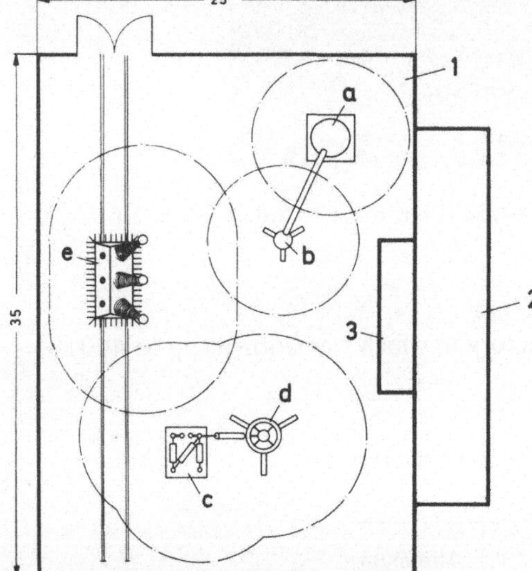

Fig. 2.1-3 Testing bay for 400 kV power transformer (dimensions in m)

1 Screened high-voltage hall

a	testing transformer	800 kV
b	capacitor	800 kV
c	impulse generator	3 MV
d	impulse voltage divider	3 MV
e	test object	

2 Adjacent testing rooms

3 Control desk

The high-voltage installations of testing bays conform to the maximum test voltage required, as well as the load represented by the test object. The guiding values listed in Table 2.1-1 are intended to give an idea of the magnitude of the test voltages for 3-phase high-voltage equipment with a given operating voltage. In any specific case, the exact values have to be taken from the appropriate test specifications valid at the time.

When selecting the voltage generators using Table 2.1-1, one should observe that the rated voltage of the generator must be chosen to be higher than the test voltage given in the table. For a.c. test transformers an increase of about 10 % of the required test voltage is sufficient. For impulse voltage generators it is usual to specify the total charging voltage, which must be multiplied by the utilization factor to derive the peak value of the impulse voltage. The utilization factor is influenced by the test object however, and, above all for the generation of switching impulse voltages, can even assume values below 0.5. The following guiding values may be given for the factor with which the highest required withstand voltage should be multiplied in order to derive the required rated voltage of the test voltage generator [*Fischer* 1969; *Leroy, Gallet* 1975].

Alternating voltage	1.1
Lightning impulse voltage	2.0
Switching impulse voltage	2.6
Direct voltage	1.7

The minimum clearances listed in the last column of Table 2.1-1 are guiding values for the lowest required air spacings between the parts of the test setup at high-voltage potential and those points in the surroundings at earth potential. At very high operating voltages the minimum clearances are determined by the magnitude of the positive switching impulse voltage to be generated, since a given electrode configuration can have a particularly low breakdown voltage for this type of voltage.

The majority of high-voltage test objects represent capacitive loading of the test voltage source. The following guiding values may be given for the capacitances [*Siemens* 1960]:

Support and suspension insulators		20 pF
Bushings, instrument transformers		200–400 pF
Power transformers		
(high-voltage winding against all other parts)	up to 1 MVA	3000 pF
	up to 100 MVA	25000 pF
Cable sample, up to 10 m long		3000 pF
Experimental setup, measuring capacitor,		
leads for a.c. test voltages	up to 100 kV	100 pF
	up to 1000 kV	1000 pF

The more common features of testing transformers are compiled in Table 2.1-2. Testing transformers are designed in most cases for short-time service of 15 minutes to 1 hour; continuous operation is necessary only for temperature-rise tests or for an investigation of dielectric stability (cables, bushings).

The corresponding data for impulse voltage generators are contained in Table 2.1-3. The usual values of the impulse capacitance lie at a few tens of nF. The table was compiled using an average value of C_s = 25 nF for the calculations. For test objects which represent particularly heavy loading (cables, power transformers), it may become necessary to choose an even larger impulse capacitance [*Widmann* 1962].

Observance of the safety regulations must be enforced with particular rigour in testing bays, since the routine nature of the work very often results in a slackening of care, even in qualified personnel. Moreover, unqualified personnel also have to enter the danger-zone, particularly during delivery and return of the test objects. Organisationally well thought out and technically reliable safety measures must prevent any risk due to the high voltage.

Further details of approved high-voltage testing bays can be found in the literature [*Elsner* 1952; *Gsodam, Stockreiter* 1965; *Läpple* 1966; *Heyne* 1969; *Raupach* 1969].

Table 2.1-2 Examples for the design data of testing transformers

Rated voltage kV	Rated current A	Rated power kVA	Short circuit impedance %
100	0.1	10	10
300	0.3	100	10
800	0.5	400	15
1200	1	1200	25
2000	2	400	25

Table 2.1-3 Examples for the design data of impulse voltage generators

Total charging voltage U_0 kV	Impulse capacitance C_s nF	Impulse energy W kWs	$\frac{W}{U_0}$ kWs/MV
200	25	0.5	2.5
400	25	2.0	5
1000	25	12.5	12.5
2000	25	50	25
4000	25	200	50

2.1.3 High-Voltage Research Laboratories

Some reference data for the selection of basic equipment for high-voltage laboratories intended for research and development work are contained in the information given for testing bays in Tables 2.1-1 to 2.1-3. It is clear from the nature of the assignment that no further details can be given about the data of technical equipment required for specific research projects. The following exposition shall be restricted to mentioning the overall aspects and to drawing attention to the description of practical setups in the literature.

Apart from meeting the technical requirements, the planning of research laboratories must also provide for the greatest possible flexibility. In the arrangement of test rooms with different dimensions and facilities, any not essential commitment to a particular application should be avoided. Besides this, it is desirable that even the largest rooms, which are designed for the highest voltages, should permit temporary subdivision in order to ensure simultaneous use. By experience the highest voltage is only rarely required.

Moreover, for high-voltage laboratories in particular, it can also be useful to provide for a multipurpose experimental hall with an average height of 6 ... 12 m and floor area of a few hundred m^2; this hall can then be subdivided into a large number of experimental stands using movable metallic barriers, depending upon the respective projects. If appropriate precautions are taken, the advantages of this kind of flexibility outweigh the disadvantages of mutual disturbances.

It is convenient if the experimental hall with the highest voltage borders on an outdoor site where unwieldy test objects in particular can be investigated, using the voltage source of the hall. Wall bushings for test voltages above 500 kV a.c. can only be produced at great expense. In addition difficulties arise with regard to safety and measuring techniques, on account of the optical separation between the voltage source and the test object, as well as due to the self-capacitance of the bushing. It would in general be better, therefore, to provide the experimental hall with a large gateway through which either the voltage sources may be temporarily taken out into the open, or through which the test voltage is led out by means of a simple wire connection.

When planning high-voltage laboratories for research and development work one should make certain that a sufficient number of auxiliary rooms are provided. Apart from the obvious rooms for offices, workshops, power supply, etc. one should above all not forget rooms for storage, stowing and packing. At a height of 3 to 6 m, these rooms should together take up at least one third of the floor area of the laboratory itself. If one economizes on this point, after a short working period it is likely that a large portion of valuable laboratory space will be blocked up by temporarily unused testing and auxiliary equipment. Of course, good transport facilities must exist between the testing rooms and the stowing areas. Coverable loading hatches in testing rooms within reach of the crane equipment and which provide access to the stowing rooms located in the cellar, have been found very useful.

Further particulars of approved high-voltage laboratories can be found in the literature [AEG 1953; *Micafil* 1963; *Prinz* 1965; *Leschanz, Oberdorfer* 1968; *Nasser, Heiszler* 1965; *Leroy* et al. 1971].

2.1.4 Auxiliary Facilities for Large Laboratories

As auxiliary facilities we understand here crane equipment, means of transport, heating, illumination etc. For testing bays in the production lines of factory halls, the choice of most of these auxiliary facilities would already be made through the mode of production.

The selection of crane equipment is important; this can take the form of either bridge cranes, overhead tackles or fixed lifting gear. For loads over 50 t only bridge cranes can be considered from the constructional point of view. These can indeed reach any part of the hall, but introduce the inherent disadvantage that suspension of elements of the experimental setup from the ceiling impedes the movement of the crane bridge. This suspension is particularly convenient for voltage carrying parts, since otherwise one would be forced to use support insulators, insulated stands or insulating wallspacers to set up the experimental circuit. This disadvantage does not arise for the approved arrangement of several overhead tackles for medium loads (2 ... 10 t). These allow experimental elements to be fastened to the ceiling by means of hooks or fixed pulleys, and can themselves be used to suspend certain components. An insulated suspension for light components, such as the high-voltage connections, can be effected with plastic ropes; for heavier loads glass-fibre reinforced plastic rods or suspension insulators are suitable.

An important decision, to be taken in view of future test objects, concerns the necessary loading capacity of the research room floors. This lies in the range of 0.5 ... 2 t/m^2, and is essentially influenced by the question whether the large-scale equipment is mobile or not. Mobility should definitely be aimed at from the standpoint of optimal utilization of the rooms. Loads over about 10 t can be placed or moved on rails embedded in the floor.

Since the freedom of installation is restricted by permanent rails, preference should normally be given to the rather more expensive solution without fixed rails. Here for heavy loads air-cushioned foundations instead of steerable undercarriages have proved their merit. Loads up to some ten tons can also be mobile at minimum expense by means of rails laid openly on the floor and interchangeable flat rollers on the apparatus base.

Regarding the heating of the testing rooms, the requirement is that it should function dust-free and noiselessly. The illumination in the testing room, which should if possible be able to be completely blacked out, must be finely adjustable, since visual observations and optical measurements represent an indispensable auxiliary aid to high-voltage experiments.

Further details of auxiliary facilities for high-voltage research laboratories can be found in the literature [*Marx* 1952; *Prinz* 1965; *Leroy, Gallet* 1975].

2.2 Fencing, Earthing and Shielding of Experimental Setups

The equipment for fencing, earthing and shielding research laboratories for high voltages is intended to prevent risk to persons, installations and apparatus. At the same time undisturbed measurement of rapidly varying phenomena should be ensured and undesirable mutual interference between the experimental setup and the environment avoided.

2.2.1 Fencing

The actual danger zone of the high-voltage circuit must be protected from unintentional entry by walls or metallic fences. Simple barrier-chains or the identification of the danger zone solely by warning signs can be considered sufficient only where their observation can be constantly supervised. Entrances to the danger zone should be provided with locks which effect automatic switch-off.

Visible metallic connection with earth must be established before the high-voltage elements are touched. In the case of smaller experimental setups, such as a practical, this can be done before entering the setup with the help of insulated rods introduced through the fencing mesh, and which establish the ground connection inside. In larger setups placing the earthing rod should be the first action after entering the danger zone, or automatic earthing switches should be provided. Complete earthing is especially important when the circuit contains capacitors charged by direct voltage.

Further particulars concerning this topic may be found in Appendix 1 under Safety Regulations.

2.2.2 Earthing Equipment

Apart from the obvious measures to guarantee reliable earth connections for steady working conditions, one must remember that rapid voltage and current variations can occur during high-voltage experiments as a result of breakdown processes. In consequence, transient currents appear in the earth connections and these can cause potential differences of the same order of magnitude as the applied test voltages.

Elements at earth potential during steady operation can temporarily acquire a high potential, though in general personal risk is not the consequence. On the other hand, damage to equipment and disturbance of measurements often takes place. The reasons for these phenomena and measures to supress them shall be discussed briefly in the following [*Stephanides* 1959; *Möller* 1965; *Sirait* 1967; *Hylten-Cavallius, Giao* 1969; *Lührmann* 1973].

The sudden voltage collapse on breakdown discharge occurs in such short times, that even lightning impulse voltages appear to be slow by comparison. The discharge which develops at the breakdown discharge site is, to a first approximation, fed by discharge of a capacitor, which, in the case of an impulse voltage generator is essentially the load capacitance and for a testing transformer the capacitances of the high-voltage winding and the test setup. Significant properties of this state of affairs can be inferred from the simple scheme of Fig. 2.2-1. A capacitor C_a begins to discharge at $t = 0$ through a gap which can be bridged in a very short time. The electrical behaviour of the circuit can be described by means of the equivalent circuit shown, where L_a denotes the inductance of the entire chopping circuit. The indicated periodically damped form of the capacitor voltage U_c and current i_a results. The first immediately recognizable requirement is that the path of the current i_a should be closed in an appropriate circuit.

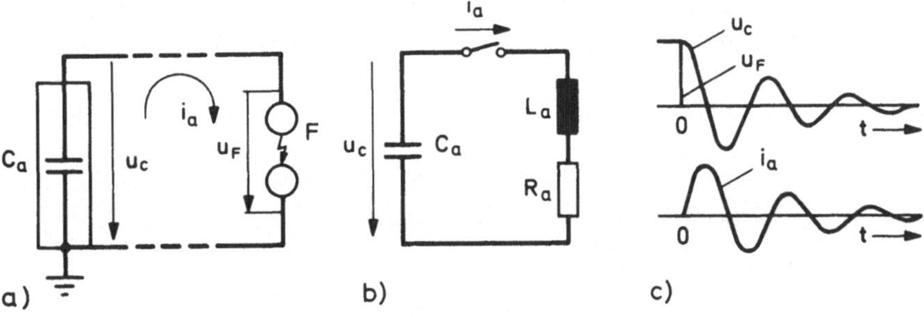

Fig. 2.2-1 Simple scheme of a high-voltage circuit with breakdown discharge process
a) Circuit setup b) Equivalent circuit c) Voltage and current curves

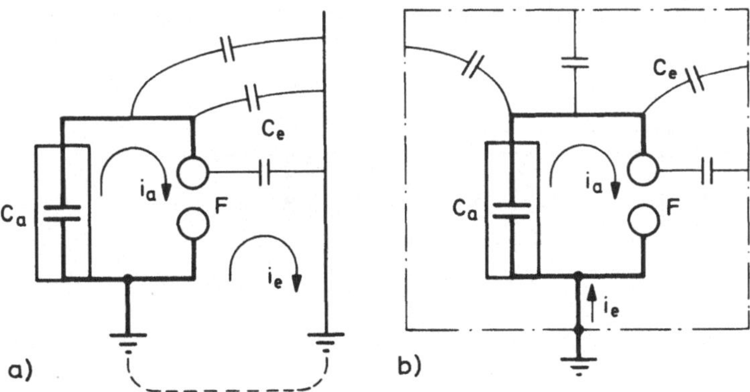

Fig. 2.2-2 Earth currents in high-voltage setups
a) without shielding b) with a Faraday cage

An electric field develops between the elements at high-voltage potential and the neighbouring elements at earth potential. This stray earth field can be simulated by a distributed earth capacitance C_e, as shown in Fig. 2.2-2a [*Sirait* 1967]. Should C_a be rapidly discharged, a transient earth current i_e is produced, generated by the potential variation of the chopping circuit; this current flows at least in part outside the test setup and here may cause undesirable overvoltages. If on the other hand the entire high-voltage circuit is surrounded by a closed metallic shield, a Faraday cage as in Fig. 2.2-2b, then the earth currents too flow in predetermined paths and the earth connections outside the cage remain current-free. These earth connections can therefore be designed exclusively according to the requirements of adequate steady operation grounding.

As a rule the floor of the laboratory, as shown in Fig. 2.2-3, is covered, at least within the region of the high-voltage apparatus, by a plane earth conductor with as high a conductivity as possible (foil or closely meshed copper grid); the earth terminals of the apparatus are connected to it non-inductively using wide copper bands, to keep the voltage drops due to large currents at a minimum. All measuring and control cables, as well as earth connections, should be laid avoiding large loops and if possible even run underneath the plane earth conductor, e.g. in a metallic cable duct [*Kaden* 1959; *Schwab* 1972].

Particular attention should be paid to the connection of the oscilloscope if disturbances during the measurement of rapidly varying phenomena are to be avoided. The measuring signal is always transferred to the measuring device via shielded cables – coaxial measuring cables as a rule. Hereby one should however prevent currents, which do not return in the

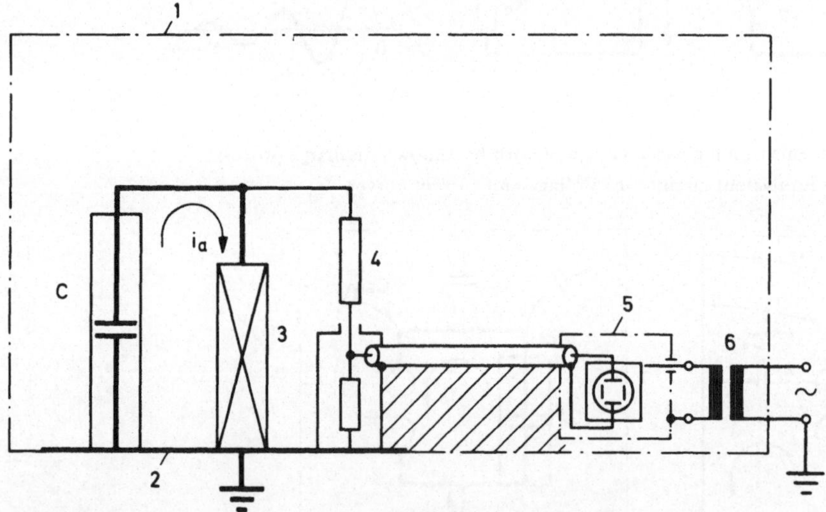

Fig. 2.2-3 Earthing and shielding of a high-voltage research setup

1. Faraday cage
2. Reinforced plane earth conductor on the floor
3. Test object
4. Voltage divider
5. KO enclosure or measuring cabin
6. Power supply to KO via isolating transformer, and low-pass filter when necessary

inner conductor, from flowing in the earthed sheath of the measuring cable, since the corresponding voltage drop is superimposed on the measuring signal as an interference voltage.

Interfering sheath currents can be prevented in various ways. If possible the highly conductive plane earth conductor on the laboratory floor is made large enough so that the cable sheath, adjacent to it and connected with it electrically at several points, is relieved of disturbing currents. Better still is the installation of the measuring cable outside the Faraday cage, e.g. in a metallic cable duct or in an earthed metal tube.

It is however often unavoidable that the measuring cable sheath and the earthing system form a closed loop in which disturbing circulating currents may flow as a result of rapidly varying magnetic fields. An earth loop of this kind is indicated by the hatching in Fig. 2.2-3 for the example of a voltage measurement. The oscilloscope enclosure in this case may only be connected by the measuring cable to the earth end of the divider and earthed; that too is why the line input of the oscilloscope has to be fed in via an isolating transformer. This isolating transformer is expedient in any case since potential differences are always present between the earth lead of the mains and the earthed parts of the experimental setup. The disturbing sheath currents may be reduced further by placing ferrite cores over the measuring cables.

In high-voltage setups with very rapid voltage variations electromagnetic waves occur, the interference signal of which may directly affect the measuring cable and the oscilloscope.

For this reason the cable and the oscilloscope used should be really well shielded. Particularly in the case of an oscilloscope with an amplifier it is advisable to set it up in a shielded measuring cabin whose line input is fed through an isolating transformer and a low-pass filter.

These same aspects which apply for the connection of oscilloscopes should be observed when direct read-out electronic peak voltage measuring devices are employed. The measurement of impulse voltages chopped on the front is particularly critical.

The peak values of the earth current $\hat{i}_e$ to be expected under unfavourable conditions rise nearly in proportion to the instantaneous value u_d of the chopped voltage. The following guiding values [*Sirait* 1967] have been obtained by experiment:

For incompletely shielded setups $\hat{i}_e/u_d \leqslant 2.5$ kA/MV
For completely shielded setups $\hat{i}_e/u_d \leqslant 6.5$ kA/MV

The peak values of i_a can be very different and usually lie well above $\hat{i}_e$. The chopping and the earthing circuit can approximately be interpreted as coupled series resonant circuits; their natural frequencies lie in the region of about 0.5 ... 4 MHz.

For a given earth current the impedance of the plane earth conductor is the decisive factor for the voltage drops produced. Further details on the calculation of these impedances can be found in Appendix 4. It is possible in this way to estimate the eventual potentials. As a guiding value it is recommended that their amplitudes be restricted to a few hundred volts [*Lührmann* 1973].

2.2.3 Shielding

Extremely sensitive measurements are often performed in high-voltage experiments. Partial discharge measurements in particular can be disturbed when the extended arrangement of the high-voltage circuit behaves as an antenna and receives external electromagnetic waves. Moreover, electromagnetic waves are also produced during breakdown discharge processes in high-voltage circuits, and these can in turn effect disturbance of the surroundings. Practice has shown that the disturbing influence of the surroundings on sensitive high-voltage measurements is generally more intense than that exerted in turn by the high-voltage investigations on the surroundings. This is mainly due to the fact that disturbing pulses occur in high-voltage circuits only occasionally and are short-lived, whereas the external disturbances, for example due to improperly screened vehicles or electric motors, generate permanent interference.

Almost complete elimination of external interference and at the same time of eventual environmental influences is achieved by using unbroken metallic shielding in the form of a Faraday cage. The standard required of the plane conductor used for this is appreciably different from that set for the plane conductor in the floor of high-voltage laboratories. Whereas for conductors intended as back-circuit for the transient current the requirement of a low voltage drop is the primary consideration, high damping of the electromagnetic fields is the objective in plane conductors intended for shielding purposes. It will therefore usually be sufficient if a close-meshed metal net is hung on or set into the walls of the laboratory and the unavoidable apertures for power and communication leads are blocked for high-frequency currents with low-pass filters. When putting this into practice special attention must be paid to careful shielding of doors and windows [*Kaden* 1959; *Prinz* 1965].

The erection of a complete Faraday cage is certainly desirable in every case, but is absolutely necessary only when sensitive partial discharge measurements are intended. Experiments in a high-voltage practical course can usually be performed without exception in partially shielded setups which have only one plane earthed conductor in or on the floor of the laboratory.

2.3 Circuits for High-Voltage Experiments

The electrical circuit for high-voltage experimental setups is appropriately made up of the three circuits shown in Fig. 2.3-1, namely the power supply circuit 1, the safety circuit 2 and the high-voltage circuit 3.

As well as switchgear equipment the power supply circuit in most cases contains an element to set the desired voltage. The safety circuit prevents the switching in, or causes switch-off, of the high-voltage circuit when one of the safety circuit switches is not closed. Finally, the high-voltage circuit consists of the high-voltage generator and measuring equipment as well as the test object.

2.3.1 Power Supply and Safety Circuits

In addition to the switchgear equipment, the arrangement for setting a variable excitation voltage for the high-voltage source is an important element of the power supply circuit.

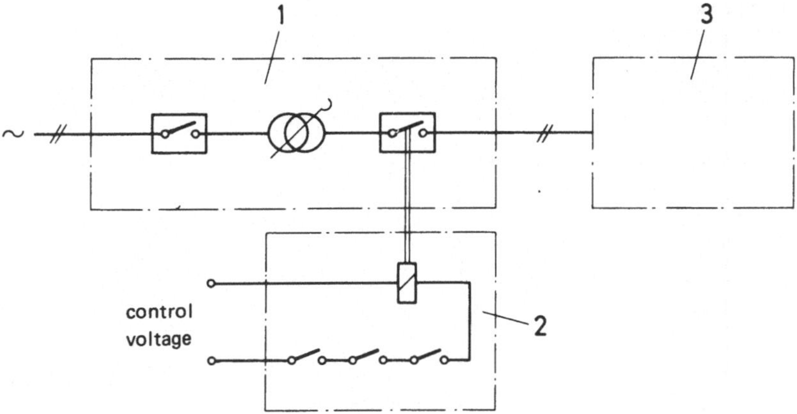

Fig. 2.3-1 Schematic representation of the basic circuit of a high-voltage experimental setup
1. Power supply circuit with regulating unit and switchgear
2. Safety and control circuits
3. High-voltage circuit with high-voltage generators, measuring equipment and test object

For power ratings up to 50 kVA, at most 100 kVA, a regulating transformer with carbon rollers is chosen. This is economical and using diverse circuits can be reliably made up to the stated high power ratings. Installation in the control desk is possible up to about 5 kVA, and manual operation is then most practicable. For higher power separate installation and remote-controlled motor drive become necessary. Above about 100 kVA a regulating transformer has to be chosen with metallic contact and load circuit breaker or excitation by means of a synchronous machine. Other possibilities are regulating transformers with rotatable winding. Whilst for excitation via a synchronous generator the supply network is relieved of power impulses during breakdown discharges in the high-voltage circuit, for excitation with regulating transformers these impulses are transferred to the network undamped.

The layout of power supply and safety circuits to be described below corresponds to that used for the practical stands in Fig. 2.1-1. The experiments are monitored from control desks which accommodate a regulating transformer for excitation of the testing transformer, as well as the most important measuring and control facilities. As protection against electrical accidents, the safety circuit effects switch-off of the testing transformer on all poles for interruption of any of the series connected safety circuit switches. The door to the test setup can only be opened with a key which also fits an interlocking switch on the control desk set in the safety circuit. The setup can only be switched on when the door of the practical stand is locked. A block diagram and the current paths of the setup are reproduced in Fig. 2.3-2 and Fig. 2.3-3. The control voltage required for energising the safety and control circuits is tapped from the supply voltage via an isolating transformer. Warning lamps mounted visibly at the entrance of the experimental stand indicate the condition of the transformer switch. Entry of the setup is only permissible

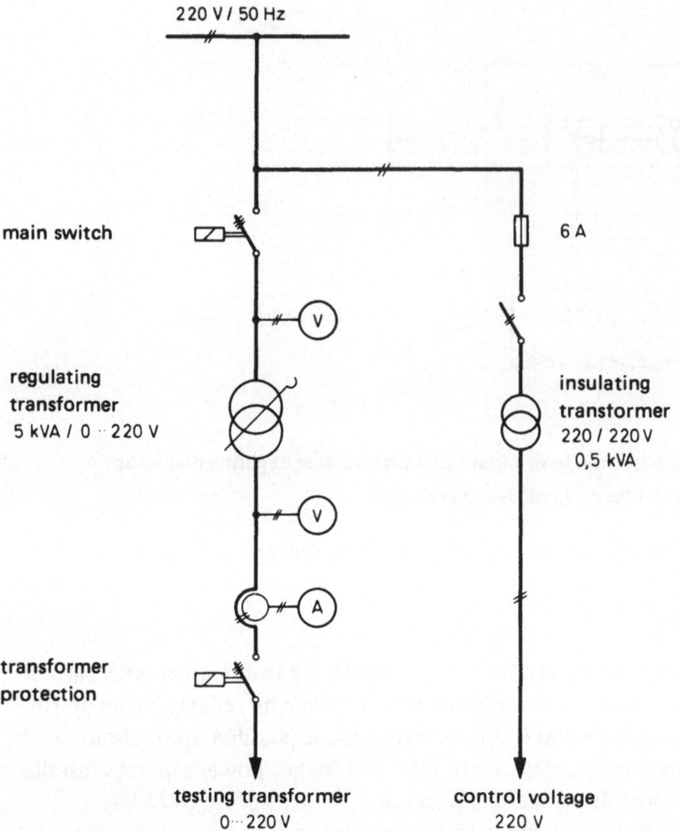

Fig. 2.3-2 Block diagram for the power supply circuit of a high-voltage experimental setup

when the green warning lamp is on. The entire illumination of the instruments, and of the warning lamps too, can be interrupted for short periods by the press-button 'Lights off', to allow observations to be made in total darkness. Set in a particularly prominent position on the control desk is the press-button 'Danger', which will immediately release the main switch when pressed.

The same holds good when the overcurrent relay indicates overloading of the power supply circuit.

2.3.2 Setting up High-Voltage Circuits

The electrical circuit diagram of high-voltage circuits is usually quite simple, since, apart from the measuring equipment, only comparatively few elements are involved. One particular difficulty, however, is that the specified clearances within the setup and to the surroundings have to be allowed for. For this reason the electrical circuit diagram alone is not sufficient for the high-voltage circuit; it usually has to be supplement by a spatial circuit diagram which indicates clearly the three-dimensional arrangement.

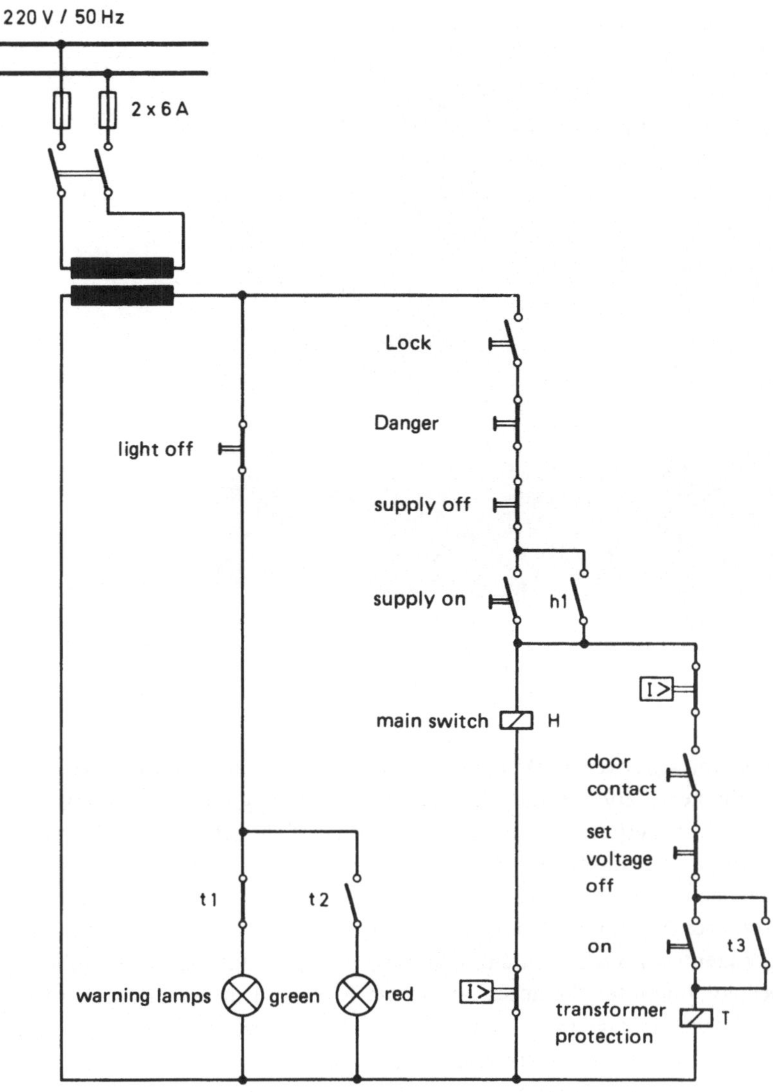

Fig. 2.3-3 Current paths of the safety and control circuits of a high-voltage experimental setup

The safety clearances given in Appendix 1 can also be used as a first estimate of the gap spacings required within a high-voltage circuit. Converted to units convenient for higher voltages, we have:

for alternating voltages	5 m for each MV
for direct voltages	3.5 m for each MV
for impulse voltages	2 m for each MV

Fig. 2.3-4 Flashover for a positive switching impulse voltage of 3.3 MV in the EHV laboratory of Electricité de France; clearance to ceiling 22 m (Photo *H. Baranger*, Paris)

For switching impulse voltages over 1 MV data of this kind cannot be given for the required spacings or the necessary minimum clearances. At strongly inhomogeneous electrode configurations, and especially for positive polarity of the electrode with smaller radius of curvature, anomalous flashovers can occur which do not take the shortest path to the opposite electrode but bridge more than double the spacing between two unforeseen points of the electrode configuration. Personnel must be completely protected by metallic barriers during experiments with high switching impulse voltages, since an adequately large clearance alone does not guarantee the necessary safety. Fig. 2.3-4 shows a flashover for a positive switching impulse voltage of 3.3 MV in an EHV research laboratory. The clearance to the ceiling was 22 m, so that a numerical value of about 7 m per MV could be deduced. But in the same setup flashovers can take place just as well to both the walls and also to the floor, although clearances between 20 m and 30 m would then be bridged.

As a rule, it is required that no predischarges may occur in a high-voltage circuit up to a certain voltage level. This requirement cannot be fulfilled by adequate clearances alone; the field strength at all points should be kept sufficiently low by appropriate curvature of the metallic parts. Whereas for voltages of at most a few hundred kV it is often still possible to mould the armatures appropriately, separate grading electrodes with large radii of curvature would be needed for very high voltages. These can also be built up of smaller component electrodes, which would considerably simplify their manufacture. When setting up high-voltage circuits care should be taken to see that all conductors which could acquire a high potential are earthed before entering the danger zone.

For this purpose safe and easily accessible connecting points for the earth leads should be provided, so that earthing can be done without risk using earthing rods.

2.4 Construction Elements for High-Voltage Circuits

Equipment for experiments with high voltage is generally set up in atmospheric air. The dimensions required of the construction elements used depend primarily upon the magnitude of the voltages appearing across them. Apart from this it is necessary to consider the dissipation of operational losses in order to avoid inadmissible overheating.

In the following several types of the more important high-voltage construction elements for indoor installation shall be briefly described. In the selection of the examples, particular priority was given to those applicable in high-voltage practical experiments and also to the feasibility of selfmade devices.

2.4.1 High-Voltage Resistors [1)]

High-voltage resistors are frequently required as charging resistors, discharge resistors or damping resistors and as measuring resistors too. Here the standard of accuracy, thermal loading capacity and dielectric strength requirements can be quite different.

Water resistors are especially suitable for applications demanding high thermal loading capacity. Rust-proof electrodes (graphite, stainless steel) are immersed in water which is usually contained in a cylinder or flexible tubing of insulating material. The value of the resistance follows from the length and cross-section of the cylindrical water container and can be varied over a wide range by using additives in distilled or tap water. A specific resistance of about 10^5 Ωcm can be attained with distilled water for longer periods; tap water takes values from 10^2 ... 10^3 Ωcm. A stability of better than ± 10 % can rarely be expected for water resistors. They are therefore only applicable where moderate demands are made on the accuracy, e.g., as charging resistors or current-limiting resistors.

In one type of construction, suitable for several purposes, a large number of low-voltage resistance elements (layer or compound type) are connected in series. In so doing it is advisable that the individual elements be arranged in such a manner that the external voltage distribution is as uniform as possible. Fig. 2.4-1 shows two designs as examples, which were developed for high-voltage practicals and correspond in their dimensions and terminal parts to the elements of the high-voltage construction kit described under 2.4.4.

In the design shown in Fig. 2.4-1a terminals are arranged between the individual resistance elements so that a voltage divider with finely variable steps is produced. To increase the permissible voltage stress and improve the heat dissipation of each resistance element, they may be immersed in oil as shown in Fig. 2.4-1b. With respect to the stability of the resistance value, one should take into consideration that the resistance value of high ohmic layer and compound type resistors increases appreciably as a consequence of frequent stressing by rapidly varying voltages [*Minkner* 1969].

---

1) Comprehensive description in *Marx* 1952; *Craggs, Meek* 1954; *Kuffel, Abdullah* 1970; *Schwab* 1972

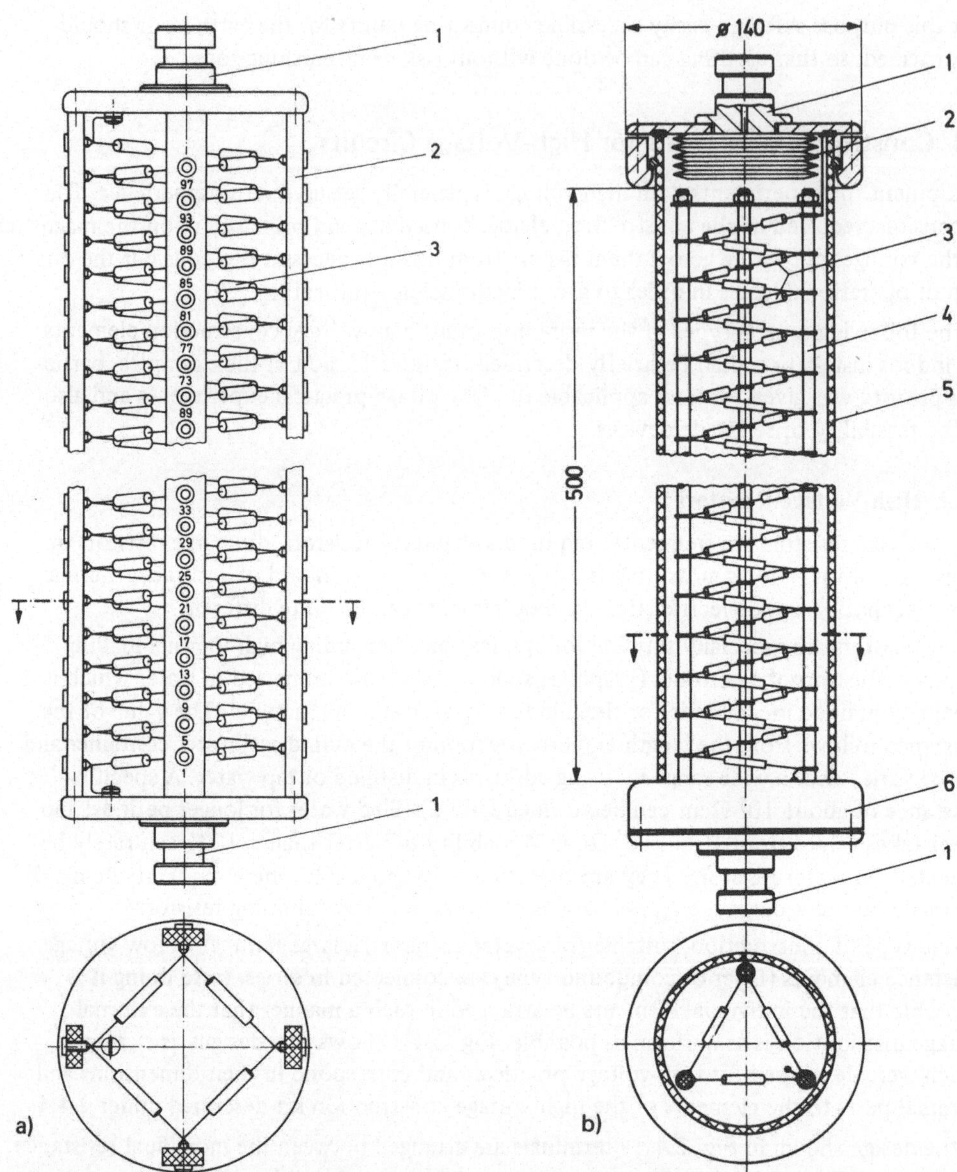

Fig. 2.4-1 Types of high-voltage resistors with carbon layer resistors

a) Measuring resistor with voltage taps in air, 25 MΩ, 140 kV, continuous service
   1. Terminal and fixing bolts
   2. Insulating material support with plug sockets
   3. Resistance element

b) Load resistor in oil-filled insulating tube, 10 MΩ, 140 kV, short-time service, 1 minute
   1. Terminal and fixing bolts
   2. Bellows
   3. Resistance elements
   4. Insulating material support
   5. Hardboard tube
   6. Metal flange

In the design of high-voltage resistors, the value of which is independent of load and time, one makes use of metallic conductors. Here the main problem is the mechanical sensitiveness of the very thin wires required for the high resistance values of, say, $10^6\ \Omega$ for each kV. Low inductance resistance mats[1] can be suspended directly in air as resistance bands, or, better still, used in special constructions, e.g. wound on an insulating support and immersed in insulating oil. Their dielectric strength reaches values of the order of 3 kV/cm; resistance values up to 6 MΩ per square metre of band area and, with self-cooled surfaces in air, continuous power ratings up to 10 kW can be realized. For particularly stringent conditions, embedding in epoxy resin has proved successful; for high resistance values mechanically stable and highly stressable resistance elements can also be produced in this way.

2.4.2 High-Voltage Capacitors[2]

In addition to resistors, capacitors are the most common elements in high-voltage circuits because they can be manufactured at more or less negligibly low loss for high voltages too. They are employed in circuits for the generation of direct and impulse voltages, as measuring capacitors and also as energy storage devices.

The dielectric of the type of capacitor usually used consists of several layers of oil-impregnated insulating paper. The thickness of the dielectric is of the order of 50 ... 100 μm. The electrodes are made of thin aluminium foil. A large number of machine-wound capacitors with partial voltages of a few kV are connected in series. For high voltages the resulting pile of capacitors is usually placed inside an oil-filled cylindrical insulating container; the capacitors CM, CS and CB of the high-voltage construction kit mentioned under 2.4.4 are made in this way. As the impregnating medium synthetic liquids or insulating gases may also be considered besides insulating oil. Very low-loss plastic foils are often used instead of the paper dielectric.

In general a ceramic dielectric has a decreasing dissipation factor with increasing frequency. It is therefore suitable for the manufacture of high-frequency capacitors for high voltages [*Hecht* 1959]. However, for technological reasons, the possible voltage per element is limited to values of about 10kV. To reach higher voltages here again a series connection of several elements must be used, which may be arranged either in air or in oil. Fig. 2.4-2 shows as an example of the application of ceramic capacitors, the construction of a damped capacitive impulse voltage divider according to 1.3.6; its dimensions and terminal parts are also adapted to the high-voltage construction kit described under 2.4.4. This divider has an extremely short response time of $T = 6$ ns, and can be used for impulse voltages up to 200 kV; the resultant capacitance is 60 pF, the resultant damping resistance 660 Ω. In order to keep the inductance of the low-voltage section as low as possible, a large number of R-C elements is connected in parallel here. The ohmic components of the high-voltage section are also made up of several parallel resistors for the same reason.

Compressed gas is suitable as a dielectric in the production of very low loss capacitors, such as those needed as the reference capacitor C_2 for the Schering bridge as in section 1.5.

---

[1] Manufacturer e.g. Schniewindt KG, Neuenrade

[2] Comprehensive treatment e.g. in *Liebscher, Held* 1968

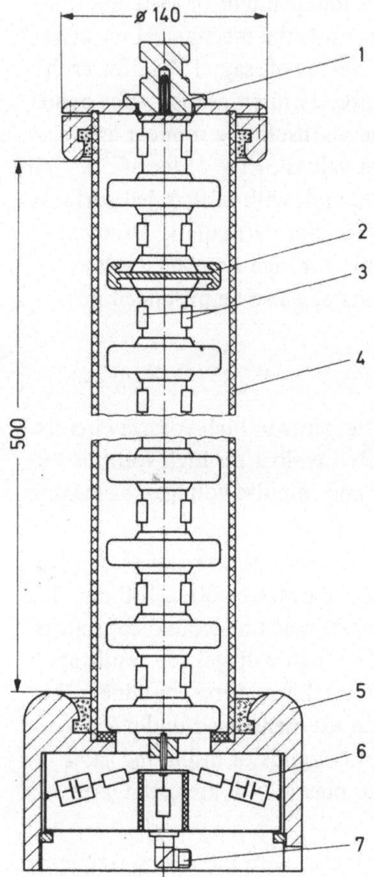

Fig. 2.4-2

Damped capacitive divider for impulse voltages up to 200 kV built up of ceramic capacitors and layer type resistors in air

1. HV terminal
2. Ceramic capacitor
3. Damping resistors
4. Insulating tube
5. Earthed metal base
6. Low-voltage section
7. Measuring cable terminal

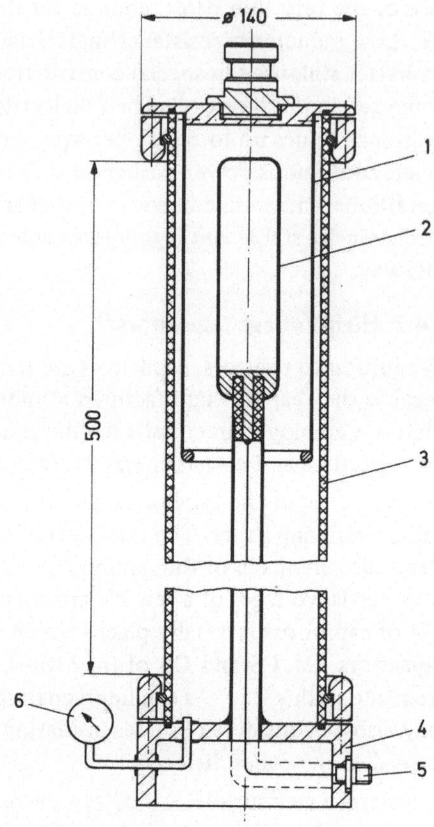

Fig. 2.4-3

Reference capacitor for 100 kV, 26 pF with compressed gas insulation

1. High-voltage electrode
2. Measuring electrode
3. Insulating tube
4. Earthed metal base
5. Measuring cable terminal
6. Manometer

The arrangement with coaxial cylinder electrodes in particular, after Schering and Vieweg, has proved to be a practical form of construction. As an example Fig. 2.4-3 shows a 100 kV compressed gas capacitor developed for the high-voltage practicals and insulated with SF_6 at 3.5 bar; its dissipation factor lies below 10^{-5}. This component too was adapted to the dimensions and terminal parts of the high-voltage construction kit described under 2.4.4. For other applications it is most important to provide an appropriate shielding electrode with a high-voltage terminal.

2.4.3 Gaps

Gaps are typical high-voltage construction elements which are used as voltage-dependent or time-dependent switches. The comparatively high resistance of the arc which establishes the conducting path between the electrodes is only rarely a disadvantage in high-voltage circuits. The gap electrodes are usually separated in the non-conducting state by a gaseous medium, preferably atmospheric air, so that repeatability of the switching process is ensured. Gaps in either liquids or solids are used only in rare cases.

Gaps with two electrodes function as voltage-dependent switching devices. They can therefore be employed as protective gaps to prevent excessive overvoltages, as switching gaps in impulse voltage circuits or as measuring gaps for voltage measurement.

A few of the most commonly used electrode configurations for 2-electrode gaps are shown in Fig. 2.4-4.

The plate-plate gap as in a) with Rogowski profile can be used to determine the breakdown voltage in a homogeneous field and is therefore suitable above all for fundamental physical investigation of the breakdown mechanism.

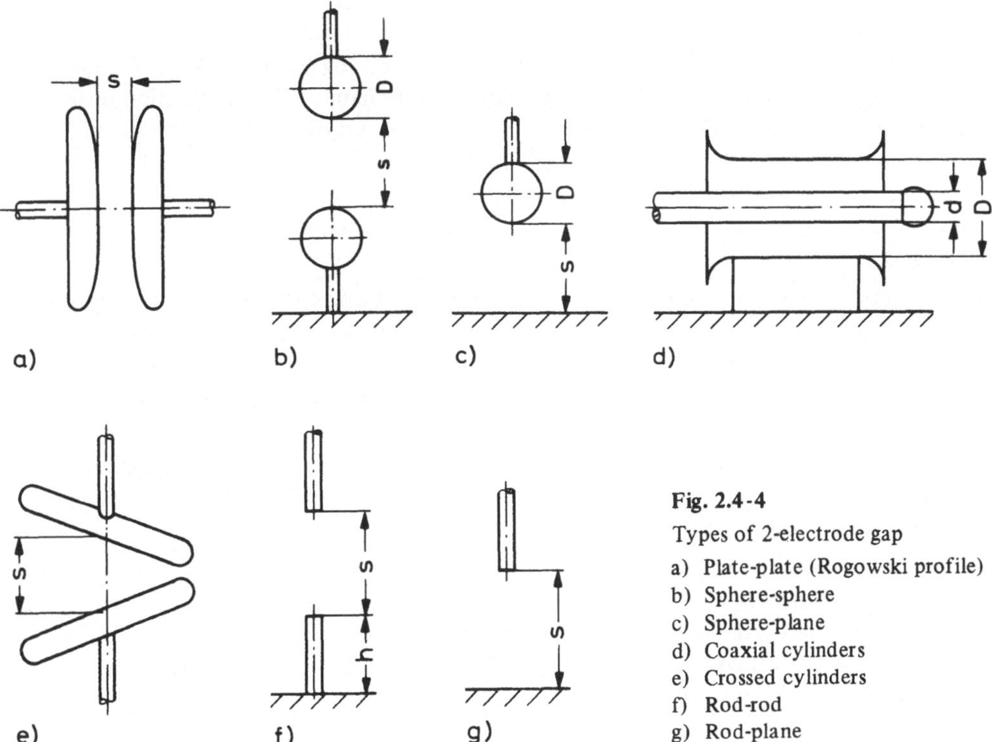

Fig. 2.4-4
Types of 2-electrode gap
a) Plate-plate (Rogowski profile)
b) Sphere-sphere
c) Sphere-plane
d) Coaxial cylinders
e) Crossed cylinders
f) Rod-rod
g) Rod-plane

The transition from a homogeneous to an inhomogeneous field can be achieved with the sphere-sphere configuration as in b) and sphere-plane configuration as in c) by variation of only one parameter, namely the spacing s. For this reason these configurations also lend themselves to basic physical investigation.

Sphere gaps for the measurement of high voltages have already been discussed in detail under 1.1.6.

The field of the coaxial cylinder gap as in d) can be calculated very accurately and the effect of edge fields eliminated by a design with guard-ring electrodes. Coaxial cylinder gaps are used in particular for investigation of the incomplete breakdown discharge at wire electrodes; this has great practical significance with regard to corona discharges on transmission lines.

Crossed cylinders as in e) are also suitable as measuring gaps because here an almost linear relationship exists between the breakdown voltage and the spacing for appreciably larger values of s/d than in the case of spheres.

Rod gaps represent the prototype of an inhomogeneous configuration and are also often used as primitive overvoltage protective devices in 3-phase high-voltage networks. The electrodes are sharp-edged rods of 100 mm^2 cross-section. It has become evident that the behaviour of the rod-rod configuration as in f) and the rod-plane configuration as in g) corresponds quite well to that of comparable electrode arrangements in practical high-voltage laboratories.

Rod gaps can also be used for measuring alternating, direct and lightning impulse voltages at a spacing from about 300 mm upwards, where the advantage of an approximately linear relationship between breakdown discharge voltage and spacing follows [*Wellauer* 1954; *Roth* 1959]. For switching impulse voltages on the other hand, the configuration positive rod against plate in particular leads to abnormally low breakdown voltages. By gradual reduction of the height h of the earthed rod in the rod-rod configuration the electrical behaviour changes to that of a rod-plane configuration, and this is of consequence to investigations of the polarity effect and the breakdown discharge mechanism of large air clearances.

Gaps with only a weak inhomogeneous field can be extended to function as time-dependent switching devices by the introduction of an auxiliary electrode; these would then, in certain ranges, be independent of the voltage between the main electrodes [e.g. *Deutsch* 1964]. To this end a voltage pulse is applied between the auxiliary electrode and one of the main electrodes at the desired instant, and this initiates the discharge between the main electrodes. This property is exploited in the time-controlled triggering of impulse voltage circuits, in chopping impulse voltages or in the simultaneous triggering of parallel impulse current circuits.

As an example Fig. 2.4-5 shows the setup of a 3-electrode gap for the high-voltage practical for a maximum working voltage of 140 kV. Installation of the auxiliary electrode is effected in a simple manner by the use of a commercial motor vehicle sparking-plug. The trigger pulse, usually fed through a coupling capacitor of about 100 pF, should have a peak value of at least 5 kV.

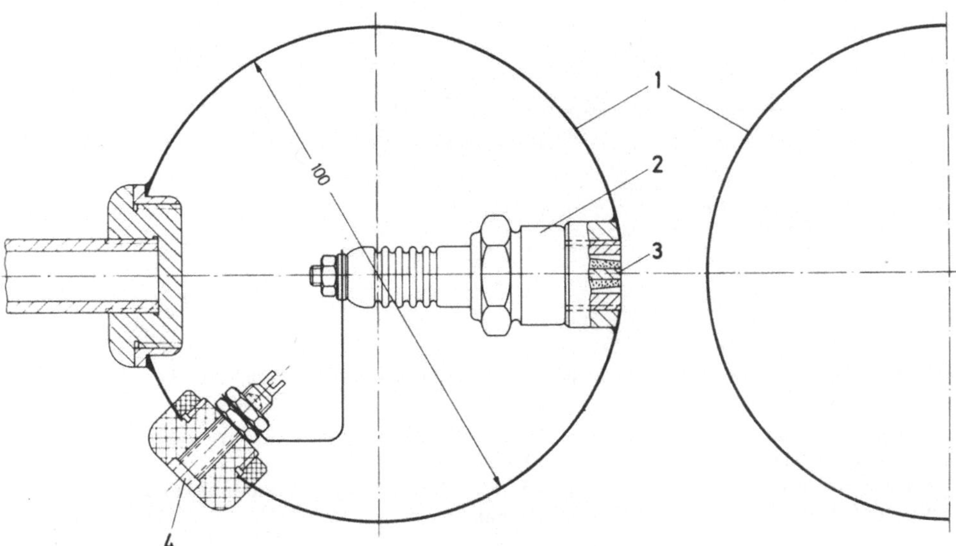

Fig. 2.4-5 Three-electrode gap for working voltages up to 140 kV

1. Main electrodes
2. Sparking-plug
3. Trigger electrode
4. Terminal for trigger pulse

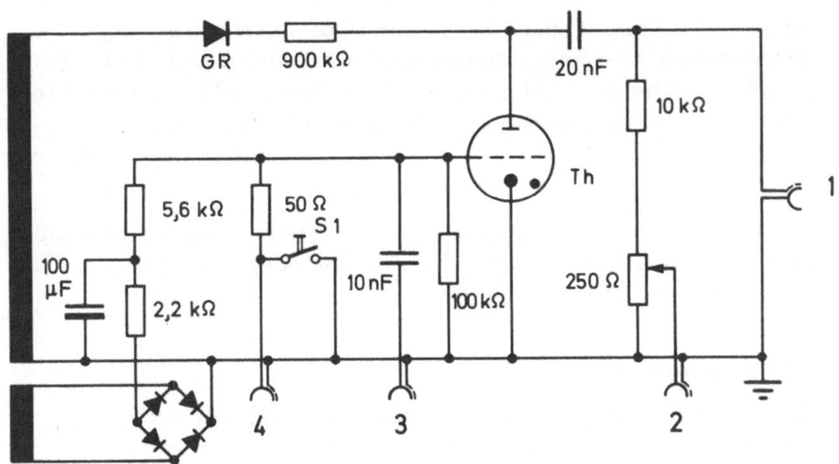

Fig. 2.4-6 Circuit diagram of a trigger device for control of impulse voltage circuits

A simple circuit for a trigger device for control of impulse voltage circuits is shown in Fig. 2.4-6. After firing the thyratron Th a negative voltage pulse appears at the output terminal 1 which is fed to a three-electrode gap either directly or via a delay cable.

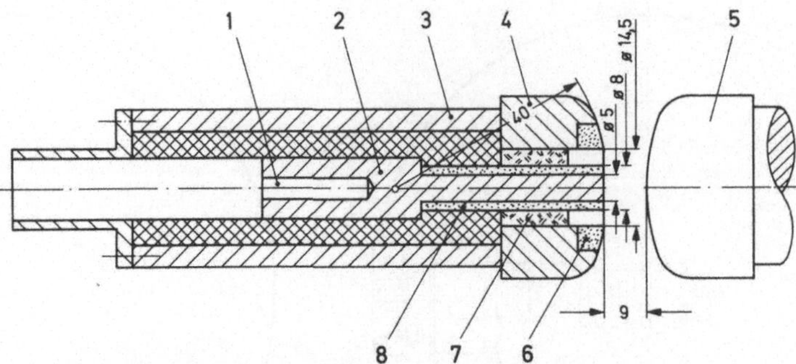

Fig. 2.4-7 Three-electrode gap for high current capacity and low firing time-lag in the voltage range of 15 to 20 kV

1. Firing cable terminal
2. Trigger electrode
3. Current circuit terminal
4. Electrode head
5. Opposite electrode
6. Hard metal sleeving (radius of curvature of the bore edge r = 0.6 mm)
7. Centering for the trigger electrode (PTFE)
8. Ceramic

To trigger an oscilloscope a pulse can be taken from the output terminal 2. Firing the thyratron can be initiated by externally shorting terminal 4, by pressing the internal closing contact S1 or by applying a positive voltage pulse to terminal 3.

The 3-electrode gap with a static breakdown voltage of 25 kV shown in Fig. 2.4-7 was developed for simultaneous firing of energy storing capacitors in the voltage range of 15 ... 20 kV. For firing impulse voltages with rate of rise of about 1 kV/ns, the firing time-lag lies below 50 ns [*Petersen* 1965]. These small time-lags have been measured for the given dimensions of the trigger electrode and its insulation. In order to guarantee that these dimensions are unchanged, even after a large number of discharges, the material of the trigger electrode as well as its insulation must possess high burn-off resistance.

For greater demands on the current capacity, temporal spread of the triggering instant and working range, different types of switching gaps and triggering systems must be chosen, each adapted to the respective application.

2.4.4 High-Voltage Construction Kit

In connection with the development of facilities for the High-Voltage Institute of the Technical University of Munich, a system of construction elements was developed by *H. Prinz* and his coworkers in collaboration with industry. This system allows orderly and rapid setting up of circuits for the high-voltage practical experiments, because of the identity of the external dimensions [*Prinz, Zaengl* 1960]. Meanwhile the system has been further improved. The experiments described in chapter 3 can of course be conducted with any arbitrary type of high-voltage element. Nevertheless, the components of the high-voltage construction kit [1]) mentioned, which are preferably used in the high-voltage

1) Manufacturer: Meßwandler-Bau GmbH, Bamberg

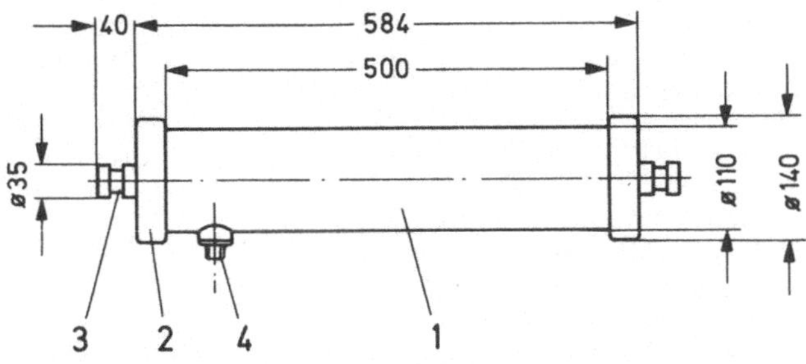

Fig. 2.4-8
Basic element of the high-voltage construction kit

1. Insulating tube
2. Metal flange
3. Terminal and fixing bolt
4. Measuring terminal for coaxial cable

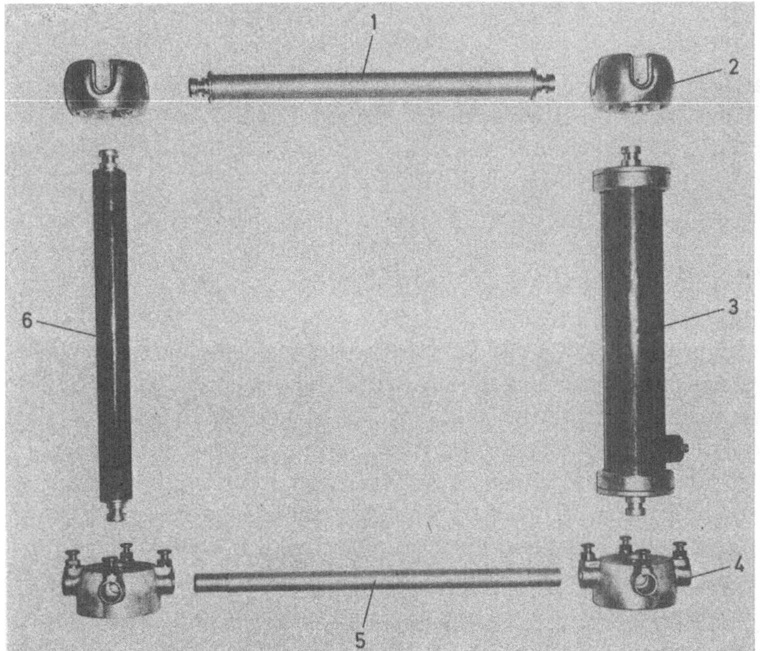

Fig. 2.4-9 Auxiliary elements for the mechanical and electrical installation of a basic element of the high-voltage kit

1. Conductive connecting rod between the connecting cups
2. Connecting cup
3. Basic element
4. Floor pedestal
5. Conductive spacer bar for insertion into the floor pedestal
6. Insulating support

practicals at the Technical University of Braunschweig, shall be briefly described here as a supplement to the literature.

The dimensions of a basic element for 100 kV alternating voltage or 140 kV direct and impulse voltage are given in Fig. 2.4-8. It has a hardboard tube casing of 110 mm external

Table 2.4-1 Basic elements of the high-voltage kit

Symbol	Construction element	Working data
CM	Measuring capacitor	a.c. 100 kV, 100 pF
RM	Measuring resistor	d.c. 140 kV, 140 MΩ
GR	Selenium rectifier	Peak inverse voltage 140 kV, 5 mA
RS	Protective resistor	Impulse, d.c. 140 kV, 10 MΩ, 60 W
CS	Impulse capacitor	Impulse, d.c. 140 kV, 6 nF
CB	Load capacitor	Impulse d.c. 140 kV, 1.2 nF
RD	Damping resistor	Impulse 140 kV, 400 Ω (1.2/50) 830 Ω (1.2/5)
RE	Discharge resistor	Impulse 140 kV, 9500 Ω (1.2/50) 485 Ω (1.2/5)

diameter; the insulating length is 500 mm. The aluminium flanges at the ends carry a cylindrical terminal bolt with an annular groove for insertion into or mounting on an appropriate support. Fig. 2.4-9 shows how a basic element can be connected to auxiliary elements required for mechanical and electrical assembly of a high-voltage circuit. The terminal and fixing bolt for assembling the components of the high-voltage kit can be unscrewed so that the component may be used in another application. The components of the high-voltage kit listed in Table 2.4-1 are used for the high-voltage practical experiments described in chapter 3.

The construction elements CM, RM, CB and RE have a measuring terminal for coaxial cables attached at the side of the insulating tube, shown in Fig. 2.4-8; this should be shorted when not in use. In addition to the basic elements, other components are available in the high-voltage kit, which have the same terminal dimensions as the basic elements (insulating support JS, sphere gap KF, conducting link V) or at least conform in their external dimensions to the kit system, e.g. 100 kV testing transformers, measuring gaps, insulated containers for investigations up to 6 bar, as well as a compressed gas capacitor as a reference for dissipation factor measurements.

Finally, a Greinacher doubler-circuit according to Fig. 1.2-5 is shown in Fig. 2.4-10, first as a three dimensional circuit diagram indicating the spatial arrangement of the individual elements and then as a photograph of the final circuit.

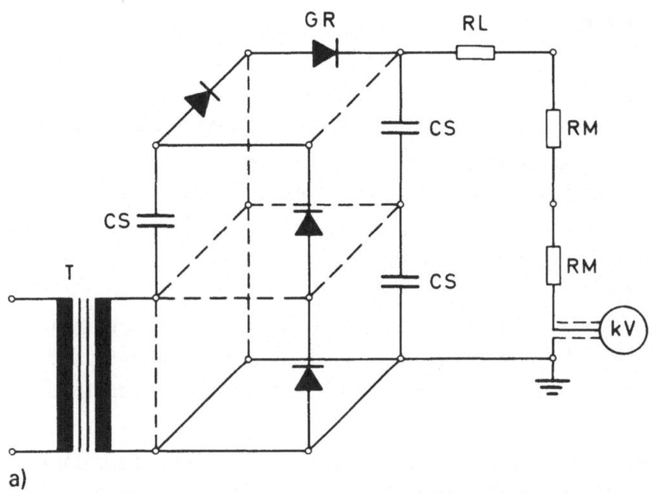

b)

Fig. 2.4-10 Setup of a Greinacher doubler circuit for 280 kV, 5 mA using the high-voltage kit
a) Spatial circuit diagram
b) Final circuit (Photo: Meßwandler-Bau, Bamberg)

3 High-Voltage Practicals

In high-voltage practicals, the students are given the opportunity to extend their theoretical knowledge obtained in the lectures by conducting experiments on their own. The experiments can be performed either in small groups of 3 to 6 participants under the guidance of a supervisor, or as joint experiments for a large number of participants. Although group experiments require a larger number of experimental setups, thereby imposing a limit on the voltage, they are to be preferred from the point of view of active participation by each member, since this is possible only in small groups. The idea of conducting experiments independently and in a responsible manner is particularly important with regard to the strict observation of safety and operational regulations, essential for the protection of personnel and equipment.

The twelve experiments described here have proved their applicability either in similar or in a slightly modified form, at the High-Voltage Institute of the Technical University of Braunschweig, as well as in other educational institutions; they are designed to be carried out in smaller groups with voltages limited to approximately 100 kV. Certain individual phenomena, which make their characteristic appearance only at very high voltages, could also be demonstrated in an additional joint experiment, if need be. The selection and performance of the experiments must at all events be undertaken with the circumstances of space, technical possibilities and personnel in mind.

The first six of the experiments given correspond to a basic practical which could be recommended as a supplement to lectures for all students of Electrical Engineering in the third academic year. The fundamental principles of experiment and questions of safety, indispensable to many other experiments, are treated in the experiment "Alternating Voltages". It is therefore strongly recommended that this one be conducted first. The sequence of the other experiments can then be chosen arbitrarily.

The next six experiments correspond to an advanced practical, which promotes the more advanced understanding and experience of those students who desire to specialize in the field of high-voltage techniques. The experiments are concerned with particular physical phenomena, with problems of measurement and testing, and with network overvoltages.

Prerequisite for meaningful participation in the practicals is thorough preparation; with each description of the experiment all those sections are mentioned which should be familiar, apart from the fundamentals inherent in that particular experiment. While conducting the experiments, the participants should preferably take turns at protocol writing, operation and reading of control and measuring instruments. Evaluation of the results should follow either immediately after the experiments, under the direction of the supervisor, or in a report to be submitted later.

The total length of an experiment should normally not exceed 3 hours. As an approximate schedule, 30 minutes could be allotted for discussion and confirmation of the necessary prior knowledge, 120 minutes for conducting the experiment proper and 30 ... 60 minutes for evaluation after the experiment. The time spent by the student working on an independent report of the experiment is generally much greater, but the advantages of this method are considerable.

Before beginning the first experiment, each participant of the high-voltage practical must confirm with his signature that he has cognizance of the safety regulations (see Appendix 1). In general the experimental circuits are to be set up by the participants, but it is essential that these be checked by the supervisor before each experiment is started. It is strictly forbidden to interfere with the safety circuits of the arrangement!

3.1 Experiment "Alternating Voltages"

Alternating voltages are required for most high-voltage tests. The investigations are performed either directly with this type of voltage, or it is used in circuits for the generation of high direct and impulse voltages.

The topics covered in this experiment fall under the following headings:

Safety arrangements
Testing transformers
Peak value measurement
RMS value measurement
Sphere gaps

It is assumed the reader is familiar with the following sections:

1.1 Generation and measurement of high alternating voltages
2.3 Circuits for high-voltage experiments

Appendix 1 Safety regulations.

3.1.1 Fundamentals

a) Equipment on the Experimental Site

The high-voltage experiments can be set up in experimental stands provided with metal barriers, as in 2.1.1. Control desks with power supply installations, safety circuits and the measuring instruments constitute the standard equipment. The control desk circuitry is shown in Fig. 2.3-2 and Fig. 2.3-3. For voltage measurement, one meter for measuring the primary voltage of the transformer and one peak voltmeter SM are provided at each desk. The change-over low-voltage capacitor of the voltage divider is incorporated in the SM.

b) Methods of Measuring High Alternating Voltages

As shown in 1.1, high alternating voltages can be measured in diverse ways. Of these, the following shall be used in this experiment:

Determination by using the breakdown voltage $\hat{U}_d$ of a sphere gap;
measurement of $\hat{U}$ with the peak voltmeter SM at the low-voltage capacitor of the voltage divider;
measurement of U_{rms} with an electrostatic instrument;
determination of $\hat{U}$ using a circuit after *Chubb* and *Fortescue*.

3.1.2 Experiment

The following circuit elements are used repeatedly during this experiment:

T Testing transformer, rated transformation ratio 220 V/100 kV, rated power 5 kVA

SM Peak voltmeter as in Fig. 1.1-10 with built-in change-over low-voltage capacitor. Connection to CM by coaxial cable

KF Sphere gap, D = 100 mm

CM Measuring capacitor, 100 pF.

a) Checking the Experimental Setup

The complete circuit diagram of the control desk and the current paths of the safety circuits (examples in Figs. 2.3-2 and 2.3-3) should be discussed and, where possible, the actual wiring of the experimental setup traced.

A series of measures which guarantee protection against electrical accidents can be identified in the circuit diagram. The faultless functioning of the safety circuit and the fulfilment of the safety regulations of Appendix 1 should be checked practically.

b) Voltage Measurements by Diverse Methods

A testing transformer T is connected as shown in Fig. 3.1-1 single phase to earth. The ratio of the secondary to the rated primary voltage is denoted by ü; a measuring capacitor CM, a sphere gap KF and an electrostatic voltmeter are connected on the high-voltage side.

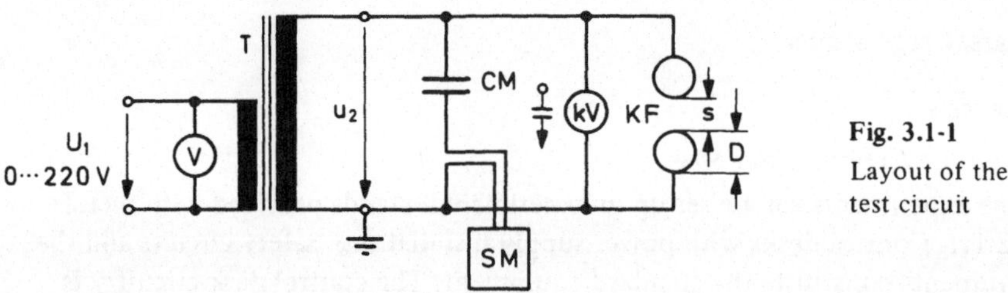

Fig. 3.1-1
Layout of the test circuit

For the gap spacings s = 10, 20, 30, 40 and 50 mm, the breakdown voltage of the sphere gap should be determined using the following methods:

$U_{2\,rms}$ by measurement (electrostatic voltmeter, directly or with a capacitive divider)

$\hat{U}_2/\sqrt{2}$ by measurement (capacitive divider with peak voltmeter SM)

$\hat{U}_d/\sqrt{2}$ from tables for sphere gaps, e.g. IEC Publ. 52, allowing for air density

$\hat{U}_2/\sqrt{2}$ according to the method of *Chubb* and *Fortescue*

For subsequent comparison, the following quantity should also be determined:

$\ddot{u}U_1$ by measurement (moving-coil instrument with rectifier at the control desk).

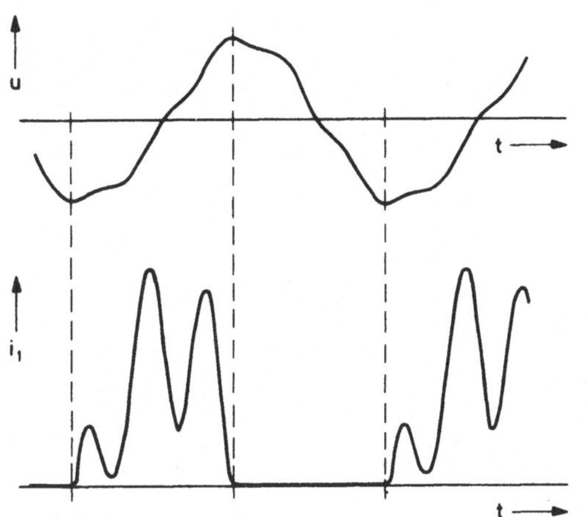

Fig. 3.1-2
Control oscillogram of the curves of the high voltage and measuring current in the Chubb and Fortescue method

The surfaces of the spheres should be polished before beginning with the measurements and several breakdowns initiated to remove any dust particles. Then five readings should be taken for each spacing and their arithmetic mean determined.

To determine the peak value by the method of *Chubb* and *Fortescue,* a device as in the circuit of Fig. 1.1-9 replaces SM. The ready-wired low-voltage part comprises two semiconductor diodes D_1 and D_2 as well as a 1 kΩ measuring resistor in the measuring branch for connection of an oscilloscope. A moving-coil instrument with 1,5 mA full-scale deflection should be used to measure the current. The curve of the voltage measured with SM should be oscillographed via the low-voltage capacitor of the capacitive voltage divider and sketched. Similarly, the curve of the current in the measuring branch should also be recorded. Thereby one should check whether the conditions imposed on the curve shape for this method are satisfied. As an example curves are shown in Fig. 3.1-2 which can be regarded as just permissible.

Note: Normally, with a testing transformer, no appreciable deviation of the voltage from the sinusoidal form will occur. For demonstration purposes, heavily distorted curves can be generated by connecting an inductance in series with the testing transformer on the low-voltage side. The non-sinusoidal magnetising current then causes a distorted voltage drop across the inductance, which in turn results in the distortion of the input voltage and with that the high-voltage output of the testing transformer.

3.1.3 Evaluation

The breakdown voltage values $\hat{U}_d$ of a spark gap, determined by the various methods of section 3.1.2b, should be represented in a diagram as a function of s. The origin of the deviation should be qualitatively explained.

Example: Fig. 3.1-3 shows the required diagram. The measured values were obtained for the comparatively heavily distorted voltage curves shown in Fig. 3.1-2. The atmospheric conditions were b = 1015 mbar and T = 296 K. The tabulated values of breakdown voltages $\hat{U}_{d_0}$ according to IEC-Publ. 52 are:

s in mm	10	20	30	40	50
$\hat{U}_{d_0}$ in kV	31,7	59	84	105	123

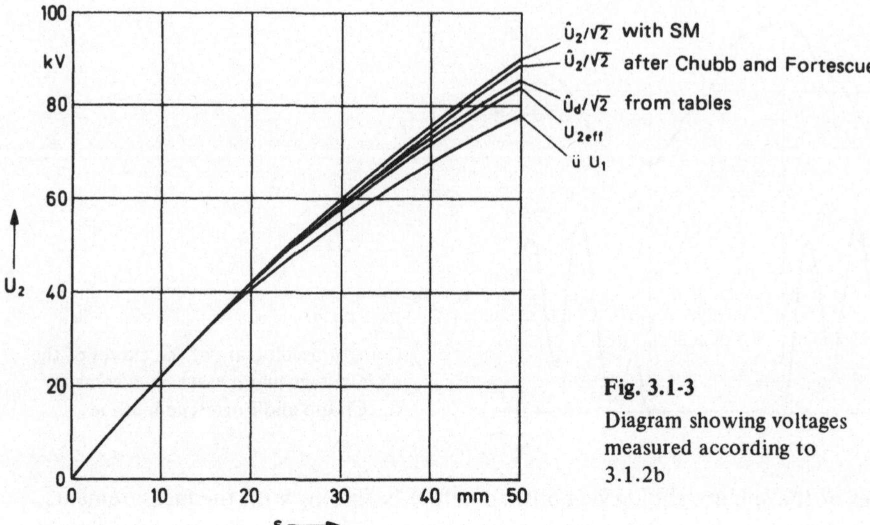

Fig. 3.1-3
Diagram showing voltages measured according to 3.1.2b

From the individually measured values of $\hat{U}_d/\sqrt{2} = d\,\hat{U}_{d_0}/\sqrt{2}$, the proportionality factor relating $\hat{U}_2/\sqrt{2}$ and $\bar{I}_1$ for the method of *Chubb* and *Fortescue* should be determined and compared with the theoretical value. The measured comparative values should also be plotted in the diagram with the appropriate characterization.

Based on Fig. 3.1-2, the time-dependent curve of i_1 should be sketched for the case of a non-smooth measuring voltage u_2. Further it should also be shown why the result would be erroneous if rectifiers were used.

Literature: *Potthoff, Widmann* 1965; *Stamm, Porzel* 1969; *Kuffel, Abdullah* 1970; *Schwab* 1972.

3.2 Experiment "Direct Voltages"

High direct voltages are necessary for testing insulation systems, for charging capacitive storage devices and for many other applications in physics and technology. The topics covered in this experiment fall under the following headings:

Rectifier characteristics
Ripple factor
Greinacher doubler-circuit
Polarity effect
Insulating screens

It is assumed the reader is familiar with section 1.2: Generation and measurement of high direct voltages.

Note: Extra care is essential in direct voltage experiments, since the high-voltage capacitors in many circuits retain their full voltage, even for a long time after disconnection. Earthing regulations are to be strictly observed. Even unused capacitors can acquire dangerous charges!

3.2.1 Fundamentals

a) Generation of High Direct Voltages

High direct voltages required for testing purposes are mostly produced from high alternating voltages by rectification and, where necessary, subsequent multiplication. An important basic circuit for this purpose is the Greinacher doubler-circuit of Fig. 1.2-5 which can at the same time be considered as the basic unit of the Greinacher cascade. The transient characteristic of this circuit when switched on can be observed in the voltage curves of Fig. 3.2-1. After switching the transformer on, the potentials of a and b increase in accordance with the capacitive voltage division, since V_2 conducts. At time t_1, V_2 stops conducting and the potential of b remains constant. The diode V_1 prevents the potential of a from falling below zero. Within the time t_2 to t_3 a current flows through V_1 which reverses the charge on C_1. At t_4 voltage division takes place once more and the entire process is repeated until the steady state condition is reached.

If a measuring capacitor is connected to the direct voltage and the alternating current through this capacitor is measured oscillographically, one can determine the ripple $u(t) - \bar{U}$ as in section 1.2.10. If a capacitive voltage divider is used, together with a peak voltmeter, its reading would then be proportional to the peak value δU. For low ripple values, the following relationship is valid:

$$\delta U \approx \bar{I}_g \frac{I}{2fC} \; .$$

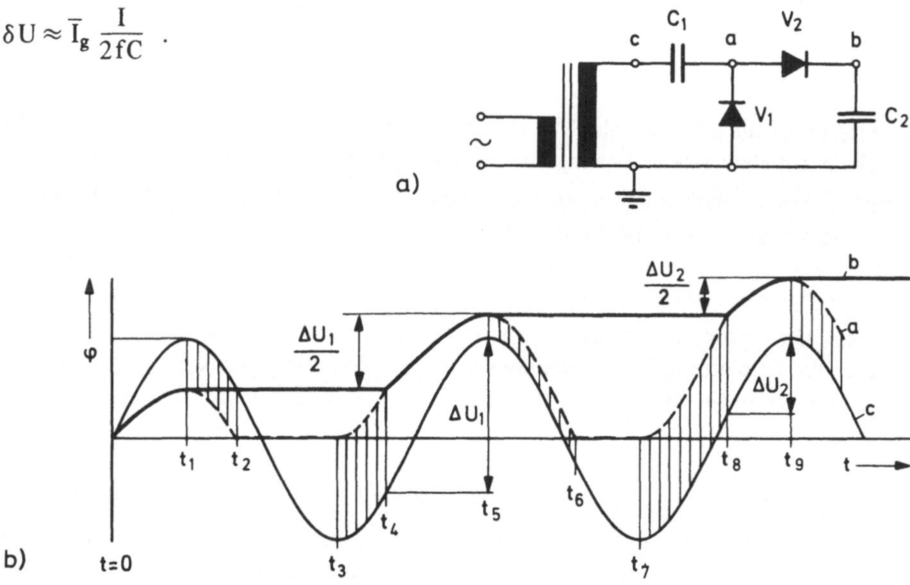

Fig. 3.2-1 Circuit diagram and voltage curves in a Greinacher doubler-circuit
a) Circuit diagram b) Voltage curves

b) Polarity Effect in a Point-Plane Gap

At an electrode with strong curvature in air, when the onset voltage is exceeded, collision ionisation results. On account of their high mobility, the electrons rapidly leave the ionis-

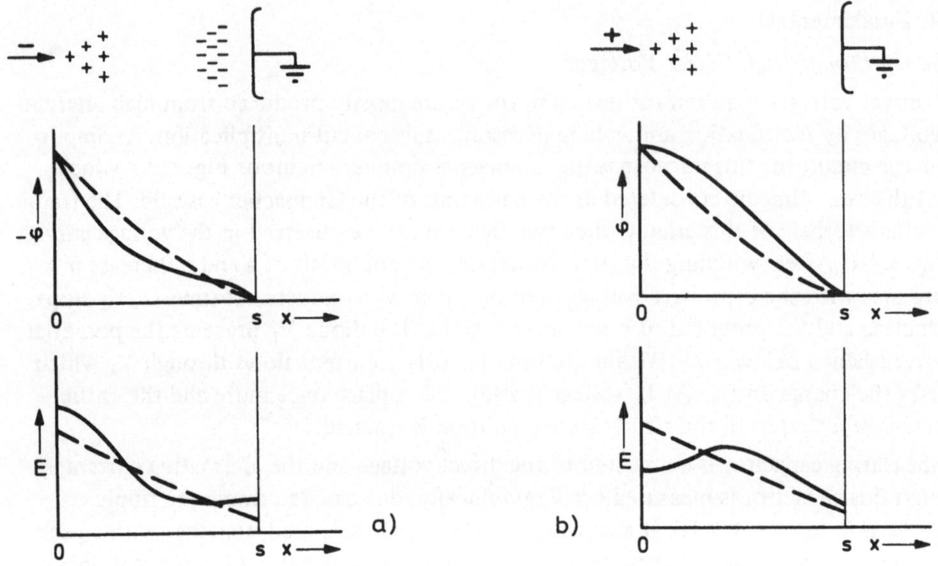

Fig. 3.2-2 Polarity effect in a point-plane gap
a) Negative point b) Positive point

ing region of the electric field. The slower ions build up a positive space charge in front of the point electrode and change the potential distribution as shown in Fig. 3.2-2.

When the point electrode is negative, the electrons move to the plate. The remaining ions cause very high field strengths immediately at the tip of the point electrode, whereas the rest of the field region shows only slight potential differences. This impedes the growth of discharge channels in the direction of the plate.

For a positive point electrode, the electrons move towards it and the remaining ions reduce the field strength immediately in front of the point electrode. Hence, since the field strength in the direction of the plate then increases, this favours the growth of discharge channels.

c) Insulating Screens in Strongly Inhomogeneous Electrode Configurations in Air

For electrode configurations in a strongly inhomogeneous field, space charges appear before complete breakdown and their distribution has a considerable influence on the breakdown voltage. Thin screens of insulating material act as a hindrance to the spreading of these space charges which cause changes in the field strength development. Because of the charge carriers accumulating on the screen, a surface charge with the same polarity as the point electrode results. Thus insulating screens, depending on their position in the field, can cause a variation of the breakdown voltage, which could, under certain circumstances, be appreciable.

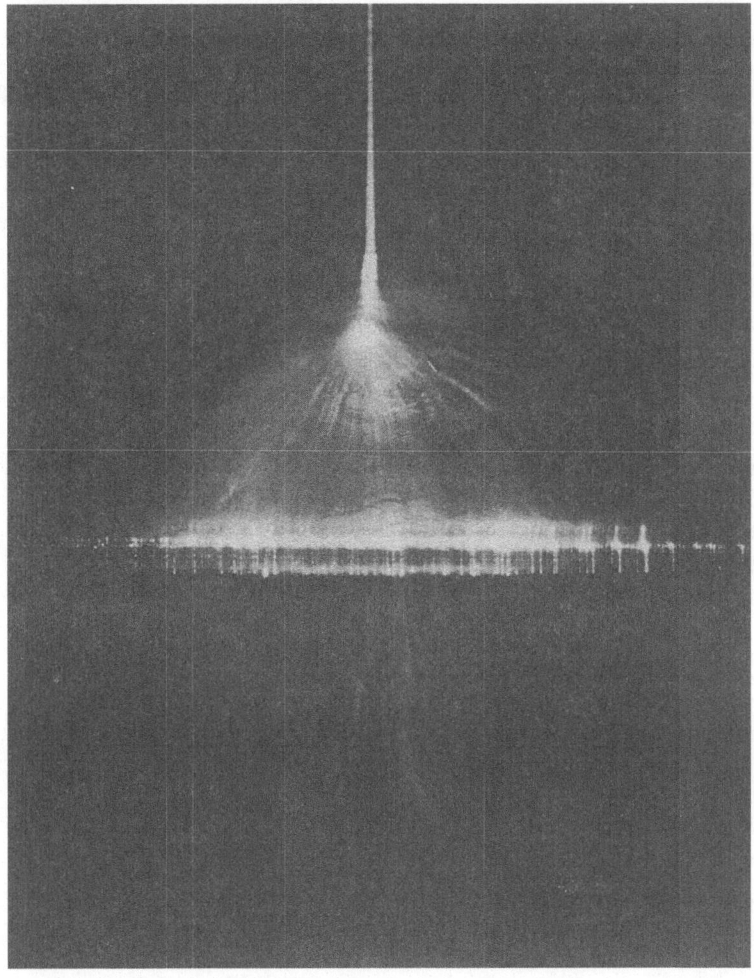

Fig. 3.2-3 Discharge photograph showing the field pattern in a point-plane gap with an insulating screen

Fig. 3.2-3 shows the result of an experiment to illustrate the effect of thin screens in an inhomogeneous field. A photographic plate was arranged along the axis of a point-plane gap. An insulating screen was placed between the electrodes, perpendicular to the photographic plate, at a distance of 9 mm from the point electrode. The figure shows the exposure on the photographic plate after a positive direct voltage of 45 kV had been applied for a few seconds.

If the screen were placed directly on one of the electrodes, it would have no effect whatsoever, since the space charge can then either build up without hindrance, or would disrupt the screen immediately. In a homogeneous field, however, a screen has no effect, since no space charges occur.

3.2.2 Experiment

For this experiment, the following circuit elements will be used repeatedly:

T Double-pole insulated testing transformer with central tap high-voltage winding, rated transformation ratio 220 V/50–100 kV, rated power 5 kVA

GR Selenium rectifier, Peak inverse voltage 140 kV, rated current 5 mA

SM Peak voltmeter (see 3.1)

GM d.c. voltmeter (moving-coil ammeter for connection to RM, 1 mA ≙ 140 kV)

RM = 140 MΩ, RS = 10 MΩ, CS = 6000 pF, CB = 1200 pF, CM = 100 pF

a) The Load Characteristics of Selenium Rectifiers

Using the components mentioned above, the circuit of Fig. 3.2-4 should be set up. The arithmetic mean value $\overline{I}_g$ of the current through the rectifiers is measured with a moving-coil ammeter in the earthing lead of T. The alternating voltage $\hat{U}/\sqrt{2}$ should be set to 50 kV. The amplitude of the direct voltage $\overline{U}$ should be measured for the following cases:

Loading only by the measurement resistance RM ($\overline{I}_g \approx 0.5$ mA)
Additional loading by RS ($\overline{I}_g \approx 5$ mA).

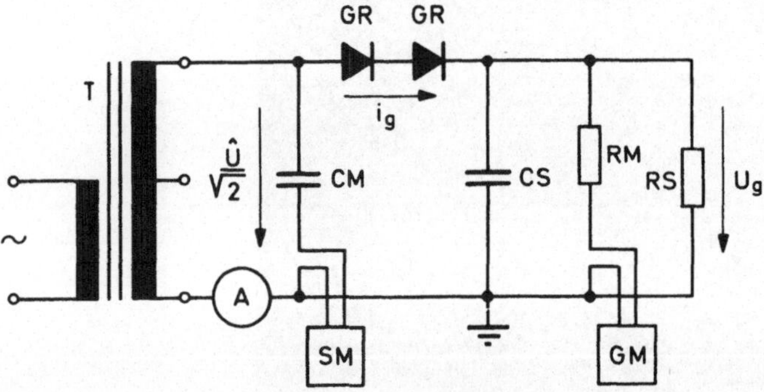

Fig. 3.2-4 Experimental setup for determining the load characteristic of selenium rectifiers

b) Determination of the Ripple Factor

The circuit can now be extended for full-wave rectification according to Fig. 3.2-5. A cathode-ray oscilloscope KO is connected parallel to the peak voltmeter. The direct current $\overline{I}_g$, as well as the peak value of the ripples δU, should be measured with the peak voltmeter SM and observed on the oscilloscope at the same time.

c) Greinacher Doubler-Circuit

The circuit in Fig. 3.2-6 should be set up. The variation in potential at point b with respect to earth is to be recorded by oscilloscope. The amplitude of the direct voltage at b, as well as the primary voltage of the transformer, should also be measured.

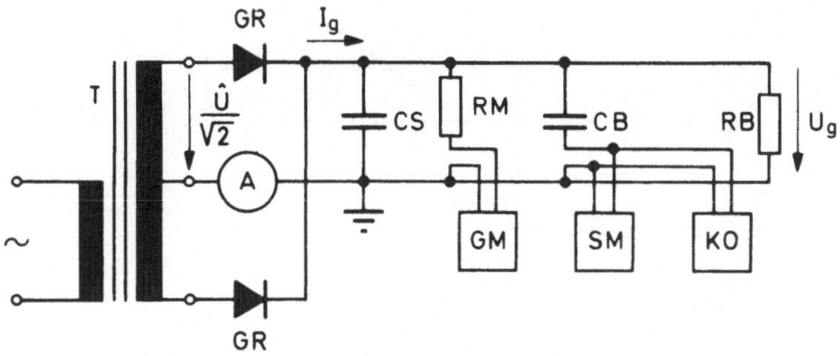

Fig. 3.2-5 Experimental setup for determining the ripple factor

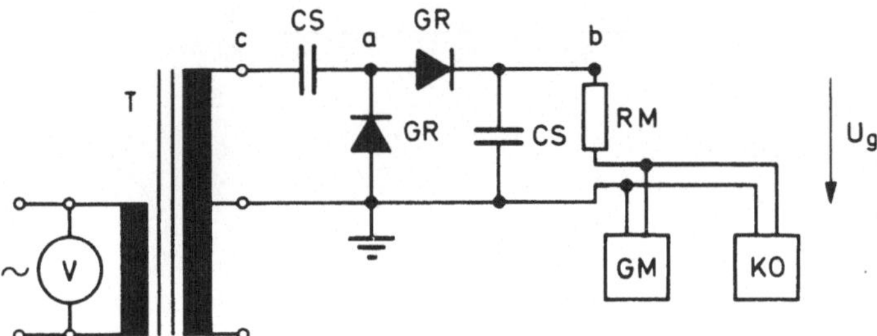

Fig. 3.2-6 Experimental setup of a Greinacher doubler-circuit

d) Polarity Effect

A point-plane gap, in series with a 10 kΩ protective resistance, is connected in parallel to the measuring resistance RM in the circuit of Fig. 3.2-6. The breakdown voltage of this spark gap should be measured for both polarities, at spacings s = 10, 20, 40, 60 and 80 mm. The transformer voltage may not be increased beyond 50 kV in this experiment, to avoid overloading of the rectifiers and capacitors.

The relationship between breakdown voltage and spacing shown in Fig. 3.2-7 was obtained for this experiment. One can see that for larger spacings and a positive point electrode, the excess positive ions in the field region lead to a lower breakdown voltage.

e) Insulating Screens

The setup is kept as in d) and the spark gap adjusted to s = 70 mm. A paper screen is held between the electrodes perpendicular to their axis, using a device for adjustment (see Fig. 3.2-3). The breakdown voltage $\overline{U}_d$ should be measured for positive point electrode, with the screen placed at x = 0, 10, 20, 40, 60 and 80 mm.

During these measurements the dependence of breakdown voltage upon the position of the screen, shown in Fig. 3.2-8, was obtained.

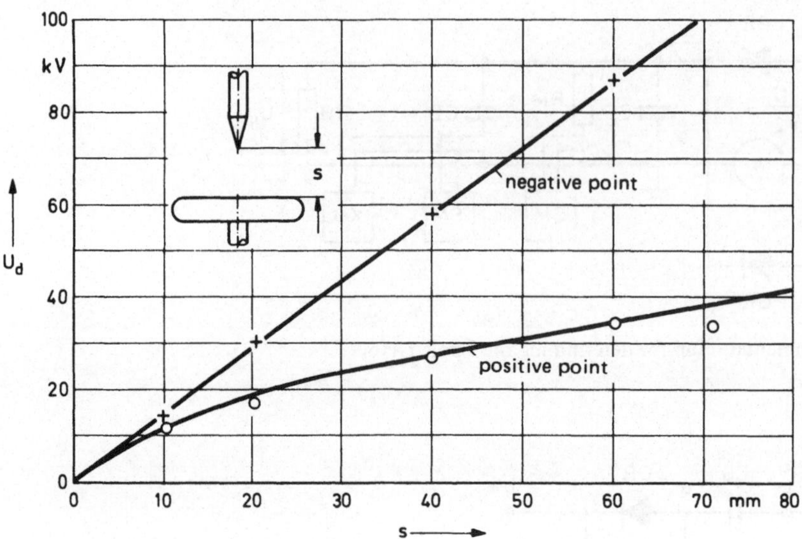

Fig. 3.2-7 Polarity effect in a point-plane gap

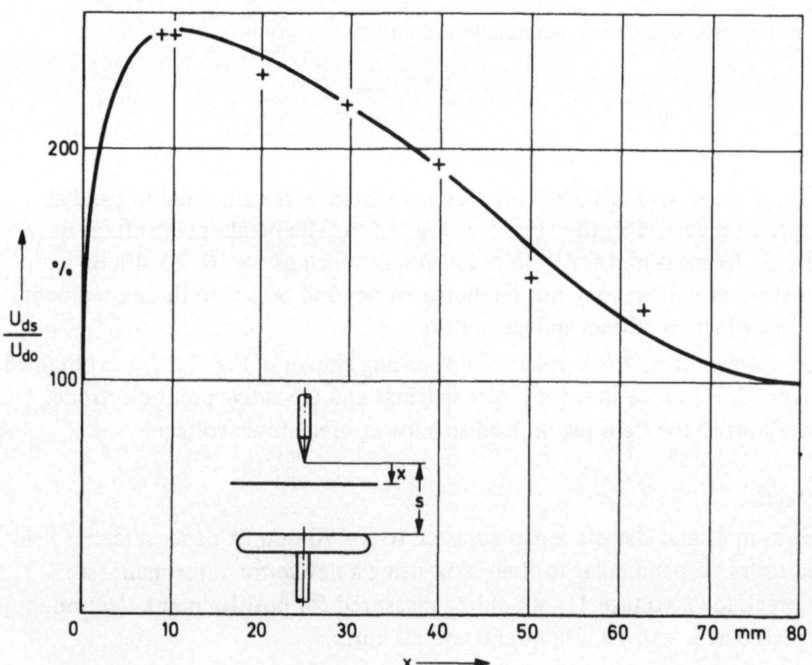

Fig. 3.2-8 Effect of thin screens on the breakdown voltage of a point-plane gap

3.2.3 Evaluation

The approximate curve of the load characteristic $\bar{U} = f(\bar{I}_g)$ as obtained under a) should be plotted. The number of series-connected rectifier plates for $U_{plate} = 0.6$ kV and the value of k should be calculated. The ripple factor measured under b) should be compared with the calculated value.

How large, in the measurements according to c), is the relative deviation of the actual direct voltage from the ideal value, calculated from the primary voltage of the transformer?

The breakdown voltages for both polarities measured under d) should be shown graphically as a function of spacing.

The value of $\bar{U}_d$ with screen, referred to the value without, should be represented as a function of x.

Literature: *Marx* 1952; *Lesch* 1959; *Roth* 1959; *Kuffel, Abdullah* 1970.

3.3 Experiment "Impulse Voltages"

High-voltage equipment must withstand the internal as well as the external overvoltages arising in practice. In order to check this requirement, the insulating systems are tested with impulse voltages. The topics covered in this experiment fall under the following headings:

Lightning impulse voltages
Single stage impulse voltage circuits
Peak value measurement with sphere gaps
Breakdown probability

It is assumed the reader is familiar with section 1.3: Generation and measurement of impulse voltages, and Appendix 5: Statistical evaluation of measured values.

3.3.1 Fundamentals

a) Generation of Impulse Voltages

The identifying time characteristics of impulse voltages are given in Fig. 1.3-2. In this experiment lightning impulse voltages with a front time $T_s = 1.2\,\mu s$ and a time to half value $T_r = 50\,\mu s$ are mostly used. This 1.2/50 form is the one commonly chosen for impulse testing purposes.

As a rule, impulse voltages are generated in either of the two basic circuits shown in Fig. 1.3-3. The relationships between the values of the circuit elements and the characteristic quantities describing the time-dependent curve were given in 1.3.3. When laying out impulse voltage circuits, one should bear in mind that the capacitance of the test object is connected parallel to C_b and hence the front time and the efficiency η in particular can be affected. This has been allowed for in the standards by way of comparatively large tolerances for T_s.

b) Breakdown Time-Lag

The breakdown in gases occurs as a consequence of an avalanche-like growth of the number of gas molecules ionised by collision. In the case of gaps in air, initiation of the discharge is effected by charge carriers which happen to be in a favourable position in the field. If, at the instant when the voltage exceeds the required ionisation onset voltage U_e, a charge carrier is not available at the appropriate place, the discharge initiation is delayed by a time referred to as the statistical time-lag t_s. Even after initiation of the first electron-avalanche, a certain time elapses, necessary for the development of the discharge channel, which is known as the formative time-lag t_a. The total breakdown time-lag, between over-stepping the value of U_e at time t_1 and the beginning of the voltage collapse at breakdown, comprises these two components, viz.:

$$t_v = t_s + t_a.$$

These relationships are shown in Fig. 3.3-1.

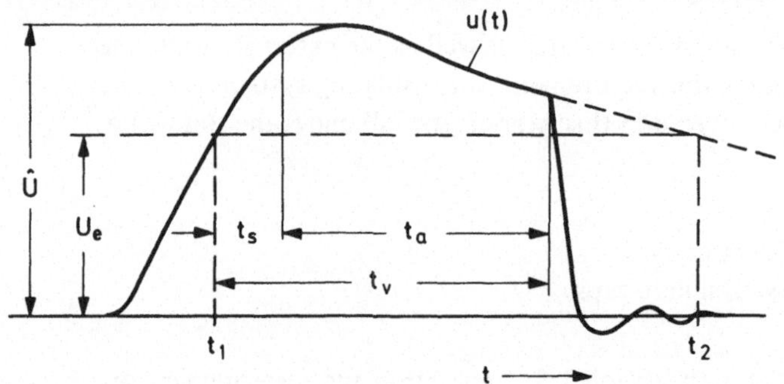

Fig. 3.3-1
Determination of break-down time-lag during an impulse voltage breakdown

c) Breakdown Probability

As a condition for breakdown one can roughly expect the time during which the test voltage exceeds U_e (Fig. 3.3-1) to be greater than the breakdown time-lag t_v. Since t_v is not constant – owing to the statistical scatter in t_s as well as some variation of t_a – repeated stressing of a spark gap with impulse voltages of constant peak amplitude $\hat{U} > U_e$ will not invariably lead to breakdown in every case. But with each mean value of the break-down time-lag one can associate an average value of breakdown voltage U_{d-50}, for which half of all the applications result in breakdown. Thus a breakdown probability P is attribut-ed to each value $\hat{U}$ of an impulse voltage of a given form. The distribution function $P(\hat{U})$ is shown in Fig. 1.3-7 for the case of a sphere gap. It is zero for $\hat{U} < U_e$ and, in the first instance, reaches a lower limiting value U_{d-0}, referred to as the "impulse withstand voltage"; knowledge of this is important for designing the insulation levels in installations. U_{d-50} is the value upon which measuring gap applications should be based. The "assured breakdown voltage" U_{d-100} represents the upper limit of the scattering region, which is of significance in protective gaps. Further information on this is contained in IEC Publ. 60-2 (1973).

Owing to the asymptotic nature of the distribution function, it is not possible to measure U_{d-0}, and especially U_{d-100}, exactly; these can, however, be determined with sufficient accuracy if the number of experiments is chosen in accordance with the width of the scattering region. Even in a series with only a few measured values, however, breakdown probabilities can be determined in an approximate manner provided a certain distribution function is assumed. Thus, using the Gaussian Normal distribution, with the arithmetic mean value U_{d-50} as well as the standard deviation s as in Appendix 5, the following approximation, approved in numerous practical cases, can be used:

$$U_{d-0} \approx U_{d-50} - 3s$$
$$U_{d-100} \approx U_{d-50} + 3s.$$

For the evaluation of such an experimental series, the measured values are usually represented on probability paper. If the plot can be approximated by a straight line, a Normal distribution may be assumed.

d) Effect of Field Configuration

For a given form of the voltage, the formative time-lag t_a is approximately constant in the homogeneous or only slightly inhomogeneous electric field of a sphere gap. Under stress of about 5 % above U_e, t_a is of the order of 0.2 μs. The breakdown probability is therefore determined primarily by the range of the statistical time-lags t_s. This can be greatly minimized by providing for charge carriers in the discharge region, e.g. by UV-irradiation. At low overvoltages, despite irradiation, the mean statistical time-lag can reach values in excess of 1 μs. Both t_a and t_s decrease very rapidly with increasing overvoltage $\hat{U}/U_e$.

The spatial as well as the temporal development of a breakdown in an inhomogeneous electric field, as in the case of a point-plane gap or in technical equipment, is different from that in a homogeneous field. Due to spatial restriction of the region in which discharge initiation can occur, the probability of a free charge carrier being there at the instant t_1 is small. The scattering zone of the breakdown voltage therefore increases at first with increasing inhomogeneity. By contrast, in such configurations where the onset voltage lies well below the breakdown voltage, charge carriers will be readily available in the electrode vicinity, so that scatter no longer occurs on account of a deficiency of charge carriers whilst the possible breakdown voltage is reached.

In a strongly inhomogeneous field, however, development of the spark channel requires a comparatively longer time than in a homogeneous field; the high charge carrier density must be transferred from the region of highest field intensity to the weaker regions; t_a also increases and is subject to considerable scatter due to the statistical nature of the spatial growth of spark channels. On the basis of these arguments it can be seen that the breakdown voltage of this kind of configuration, especially for large gaps, varies much more than that of a sphere gap for instance.

3.3.2 Experiment

The following circuit elements are used repeatedly in this experiment:

T Testing transformer, rated transformation ratio 220 V/100 kV, rated output 5 kVA
GR Selenium rectifier, PIV 140 kV, rated current 5 mA
F Trigger gap, sphere gap with trigger pin according to Fig. 2.4-5, D = 100 mm
ZG Trigger generator for generating 5 kV pulses, as in Fig. 2.4-6
UG d.c. voltmeter (moving-coil ammeter for connection to RM, 1 mA ≙ 140 kV)
KF Sphere gap, D = 100 mm.

The data of the construction elements used are as follows:

CS = 6000 pF, CB = 1200 pF, RS = 10 MΩ
RM = 140 MΩ, CM = 100 pF
For impulse voltage 1.2/50: RD = 416 Ω, RE = 9500 Ω
For impulse voltage 1.2/5: RD = 830 Ω, RE = 485 Ω

a) Investigation of a Single-Stage Impulse Generator

A single-stage impulse generator should be set up in circuit "b" as shown in Fig. 3.3-2. The voltage efficiency η of the setup should be determined at a d.c. charging voltage U_0 of about 90 kV. The peak value of the impulse voltage $\hat{U}$ should then be measured using the sphere gap KF. For this purpose, a number of voltage impulses of constant peak amplitude are applied to the sphere gap, and its spacing is varied until about half the applied impulses result in breakdown. The peak value of the impulse voltage may be determined from the gap length, allowing for the air density. This measurement should be carried out with both polarities for the 1.2/50 impulse and with negative polarity alone for the 1.2/5 impulse. Using the circuit elements provided for the 1.2/5 impulse, the voltage efficiency for circuit "a" should also be determined.

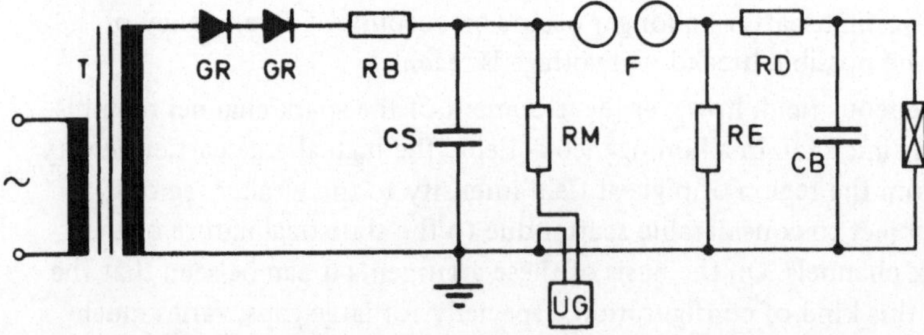

Fig. 3.3-2 Experimental setup of a single-stage impulse generator

This experiment was carried out for circuit "b" with the voltage impulse form 1.2/50, at relative air density d = 0.97, and the following results were obtained:

Charging voltage	: 90 kV
Spacing of the sphere gap	: 24.5 mm
$\hat{U}_d$ according to tables	: 70.7 kV
$\hat{U}_d$ for d = 0.97	: 68.5 kV
η	: 81.5 %
η calculated from circuit elements	: 83.3 %

b) Distribution Function of Breakdown Probability

The single-stage impulse generator should be set up as described in section 3.3.2 (a) using basic circuit "b" (Fig. 1.3-3) for generation of a positive 1.2/50 lightning surge. The trigger generator ZG, connected as in Fig. 3.3-3 to the trigger gap F (built up as a three-electrode gap) via the coupling capacitor CM, allows precise triggering of the impulse generator at an accurately preset charging voltage. One of the spheres is provided with a trigger pin to which a voltage pulse is fed through CM. The breakdown between the trigger pin and the surrounding sphere initiates the breakdown of the trigger gap.

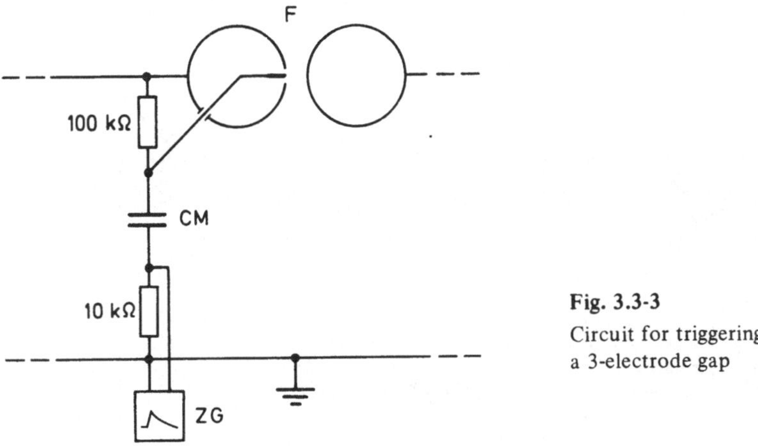

Fig. 3.3-3
Circuit for triggering a 3-electrode gap

The peak value of the impulse voltage is derived using the charging voltage U_0 and the efficiency previously determined in 3.3.2a:

$$\hat{U} = \eta \, U_0 .$$

This procedure is permissible here, since the test object capacitance is small compared with the permanently connected load capacitance CB. The chosen test objects are a 10 kV support insulator with protective gap (spacing 86 mm), representative of the inhomogeneous field configuration, and then a sphere gap (D = 100 mm, spacing 25 mm) with only a moderately inhomogeneous field. To record the distribution function, the voltage should be increased beyond the breakdown range of the test object in steps of about 1 kV, until for 10 impulses initially 0 %, finally 100 % flashovers occur.

The measured values of both test objects should be plotted on probability paper and approximated by a Normal distribution. Hence the values of U_{d-50} as well as of s, converted to standard conditions, should be determined and the values of U_{d-0} and U_{d-100} approximately obtained. From this experiment the distribution functions $P(\hat{U})$, shown in Fig. 3.3-4, were obtained. It is evident that the scatter in the breakdown voltages of the strongly inhomogeneous arrangement of the insulator is appreciably greater than for the case of the sphere gap.

Evaluation using the straight lines drawn in Fig. 3.3-4 as approximations for a Normal distribution, gave:

Sphere gap		Support	
U_{d-50}	$= 72.2\ \text{kV}$	U_{d-50}	$= 80\ \text{kV}$
s	$= 1.3\ \text{kV}$	s	$= 4.4\ \text{kV}$
$v = \dfrac{s}{U_{d-50}}$	$= 1.8\ \%$	$v = \dfrac{s}{U_{d-50}}$	$= 5.5\ \%$
U_{d-100}	$= 76.1\ \text{kV}$	U_{d-100}	$= 93.2\ \text{kV}$
U_{d-0}	$= 68.3\ \text{kV}$	U_{d-0}	$= 66.8\ \text{kV}$

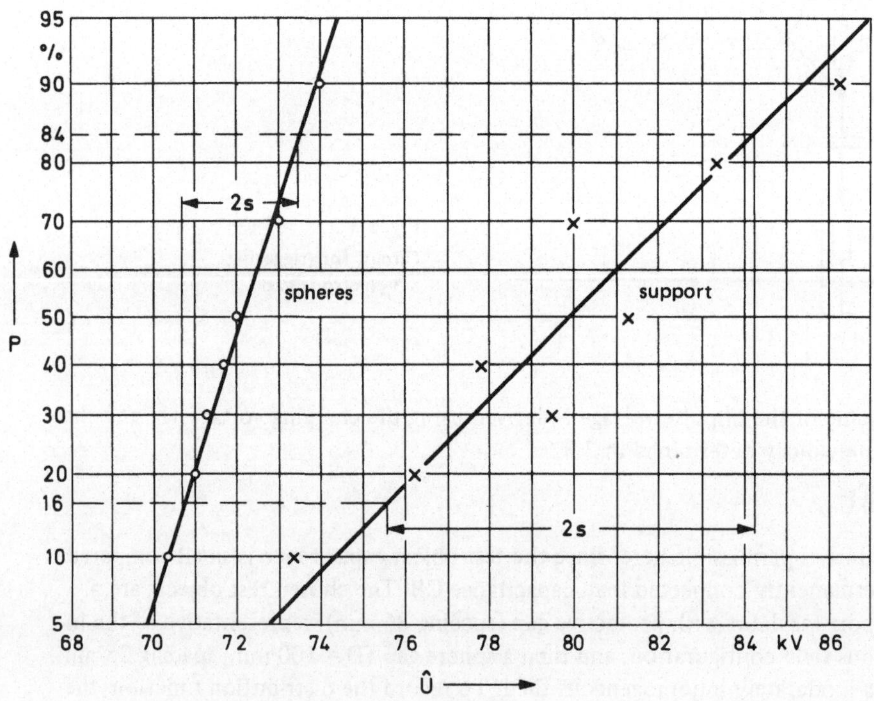

Fig. 3.3-4 Measured distribution functions of the impulse breakdown voltages of a sphere gap and a 10 kV support insulator with protective gap

3.3.3 Evaluation

The characteristic front-time T_s and time to half value T_r should be calculated from the data of the impulse circuit outlined in section 3.3.2a. The measured voltage efficiency should be compared with the theoretical value.

The relationship $P(\hat{U})$ should be determined according to 3.3.2b for the sphere gap and the support insulator and plotted on probability paper. The values of U_{d-0} and U_{d-100} should be stated for standard conditions.

The scattering ranges of both the arrangements investigated should be compared and reasons given for the differences.

Literature: *Marx* 1952; *Lesch* 1959; *Strigel* 1955; *Kuffel, Abdullah* 1970

3.4 Experiment "Electric Field"

A measure of the electric stress of a dielectric is the electric field strength, the determination of which is therefore an important task of high-voltage technology. The topics covered in this experiment fall under the following headings:

Graphical field determination
Model measurements in electric fields
Field measurements at high voltages

It is assumed the reader is familiar with the basic theory of electrostatic fields.

3.4.1 Fundamentals

By the electric strength of an insulating material one understands that value of the field strength which is just permissble under given conditions such as voltage type, stress duration, temperature or electrode curvature. The limit of electric strength of an insulating medium is reached when its disruptive field strength is exceeded at some point. For this reason the determination of the prevailing maximum field strength is of great practical significance.

An exact calculation of the electric field using Maxwell's equations [*Lautz* 1969; *Prinz* 1969], even by applying such special methods as the method of images, conformal mapping or coordinate transformation, is restricted to comparatively few, geometrically simple configurations. The introduction of computers, enabling numerical computation of even complicated field configurations, has initiated noticeable progress in high-voltage techniques. In addition, graphical and experimental methods of determining the electric field have also proved their merit.

a) Graphical Field Determination

The path of the electric field lines is determined by the direction of the electric field strength $\vec{E}$. They are orthogonal to the equipotential lines at any point and hence perpendicular to metal electrode surfaces. Under the condition that no surface charges exist in the boundary area between two dielectrics, the normal components of the field strengths are inversely proportional to the dielectric constants of the insulating materials. On the other hand, the tangential component of the electric field strength is continuous along the boundary.

For the case of two-dimensional fields, the field plot can often be obtained graphically with sufficient accuracy. The method is based upon the principle that the equipotential lines and the field lines are estimated first and then the field plot is corrected step-by-step by applying the fundamental electrostatic field laws. Those areas enclosed by adjacent field lines, as in Fig. 3.4-1, have the same electric flux $\Delta Q = b\, l\, \epsilon_r \epsilon_0 E$, where l is the extension of the configuration perpendicular to the plane of the paper and $\epsilon_r \epsilon_0 = \epsilon$ the dielectric constant of the dielectric medium. If the constant potential difference between two neighbouring equipotential lines is substituted for E, viz. $\Delta\varphi = E\,a$, the following condition results:

$$\epsilon_r \frac{b}{a} = k\,.$$

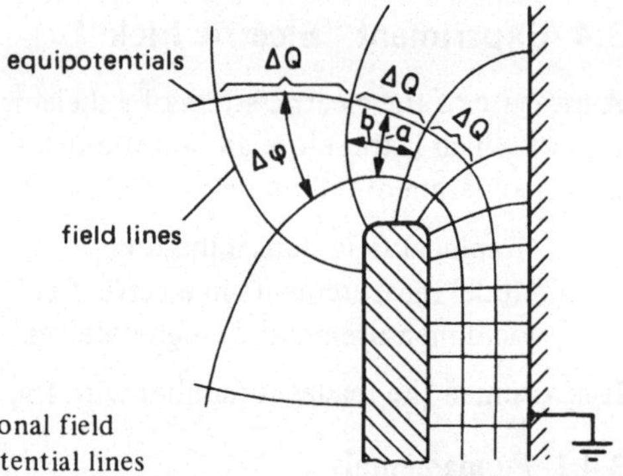

Fig. 3.4-1
Example of a two-dimensional field with field lines and equipotential lines

The constant k can be chosen arbitrarily. In the example shown, it is assumed that $b/a = 1$. Let the spacing between two neighbouring equipotential lines be a_1 at any point, then the electric field strength at that point is given by:

$$E_1 = \frac{\Delta\varphi}{a_1}\,.$$

If m is the number of equipotential lines drawn (not counting the electrode surfaces), the total applied voltage is

$$U = (m+1)\,\Delta\varphi\,.$$

If the number of field lines drawn between the electrodes is n, the total electric flux is given by:

$$Q = n\, b_1\, l\, \epsilon_0 \epsilon_r E_1\,.$$

Substituting accordingly, we have for the capacitance of the configuration:

$$C = \frac{Q}{U} = \frac{n}{m+1}\, k\, l\, \epsilon_0\,.$$

A modified version of this method may also be applied to three-dimensional fields, provided these possess rotational symmetry. An analogous argument yields the relationship:

$$\epsilon_r \frac{b}{a} r = k$$

for the reproduction of the field plot, where r represents the distance of the volume element in question from the rotational axis.
Graphical field determination can be simplified considerably if certain initial data concerning the field configuration are already available. These can be obtained either by calculation or by experimental determination of the field direction, or from the known potentials of individual points. To determine the maximum field strength, it is usually unnecessary that the entire field configuration be known, but rather the field only at positions recognised as critical.

b) Analogue Relationships between the Electrostatic Field and the Electric Field

Field measurements on models make use of the analogy between the electrostatic field and the electric field. The following relationships are valid:

Electrostatic field		Electric field
$\vec{D} = \epsilon \vec{E}$		$\vec{S} = \kappa \vec{E}$
$\iint \vec{D} \, d\vec{A} = Q$		$\iint \vec{S} \, d\vec{A} = I$
	$\vec{E} = -\operatorname{grad} \varphi$	
	$\operatorname{div} \operatorname{grad} \varphi = 0$	
$C = Q/U$		$1/R = I/U$

The distribution of field lines and equipotential lines follows the same mathematical laws in each case and depends only upon the geometry and materials. Hence the dielectric flux density $\vec{D}$ corresponds to the current density $\vec{S}$ and the dielectric constant ϵ of the electrostatic field is simulated by the specific conductivity κ of the electric field. If the ohmic resistance R of a configuration is known, the capacitance C can be calculated as:

$$C = \frac{1}{R} \frac{\epsilon}{\kappa}.$$

These analogue relationships form the basis for the application of the electrolytic tank as well as for electric field simulation with conducting papers.

c) Electrolytic Tank

A scale model of the electrode configuration is set up in a tank with insulating walls, filled with a suitable electrolyte (e.g. tap water). An alternating voltage is the appropriate choice of working voltage, to avoid the polarisation voltages arising in the case of direct voltages. The equipotential lines, or equipotential areas in the case of the electric field, are measured by means of a probe which can be fed with different voltages from a potential divider via a zero indicator.

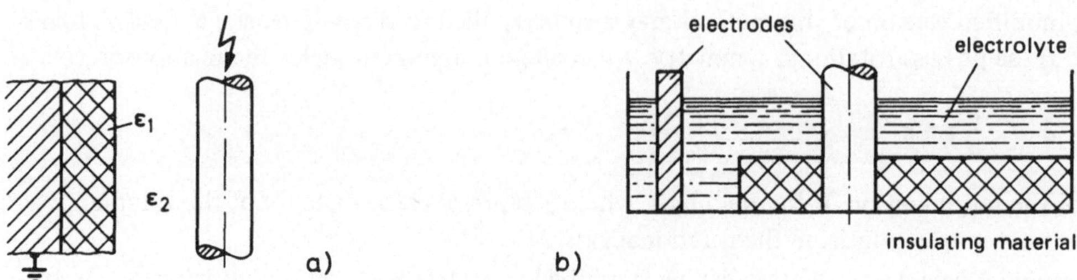

Fig. 3.4-2 Simulation of a cylinder-plane configuration in the electrolytic tank
a) Original b) Simulation for the case $\epsilon_1 = 2\epsilon_2$

Guiding the probe along the lines corresponding to the potentials selected on the divider, as well as their graphical representation, can be undertaken manually or automatically in large systems. For the two-dimensional field model, various dielectric constants can be simulated by different heights of electrolyte, as shown in Fig. 3.4-2 for a cylinder-plane configuration.

Three-dimensional fields with rotational symmetry can be readily simulated in a wedge-shaped tank, whereas for fields with no rotational symmetry one has to resort to much more complicated forms of three-dimensional simulation.

d) Simulation of Electric Fields with Conducting Paper

Two-dimensional fields can be measured easily, and also in most cases with adequate accuracy, by means of conducting paper; the number of layers of paper arranged one above the other must be chosen to be proportional to the dielectric constant at the respective position. Graphite paper has been found useful as conducting paper, having a surface resistance (viz.resistance measured between the opposite sides of a square sample) of about 10 kΩ, similar to that commonly used as conducting layer in high-voltage cables. The electrode surfaces are represented by conducting silver paint, by fixed metal foils, by electrically connected needles or spikes driven into a wooden base, or by impressed metal objects. At the boundary surfaces between electrodes and dielectrics, or between two different dielectrics, the conducting paper layers must be electrically well-connected with each other. Pins driven into the base are particularly well-suited for this purpose.
A distinct advantage of this method is that the field plot can be directly drawn onto the paper; transfer of the measured values and representation in a separate drawing are thus unnecessary. Basically, this method is also suitable for measuring three-dimensional fields with rotational symmetry; the number of paper layers must then be increased in proportion to the distance from the rotational axis.

e) High-Voltage Field Measurements

The direction of the field strength at various points of a configuration in air, as well as the potential of these points, can be determined by high-voltage measurements. Photographic record of predischarges by chopped impulse voltages can provide information concerning

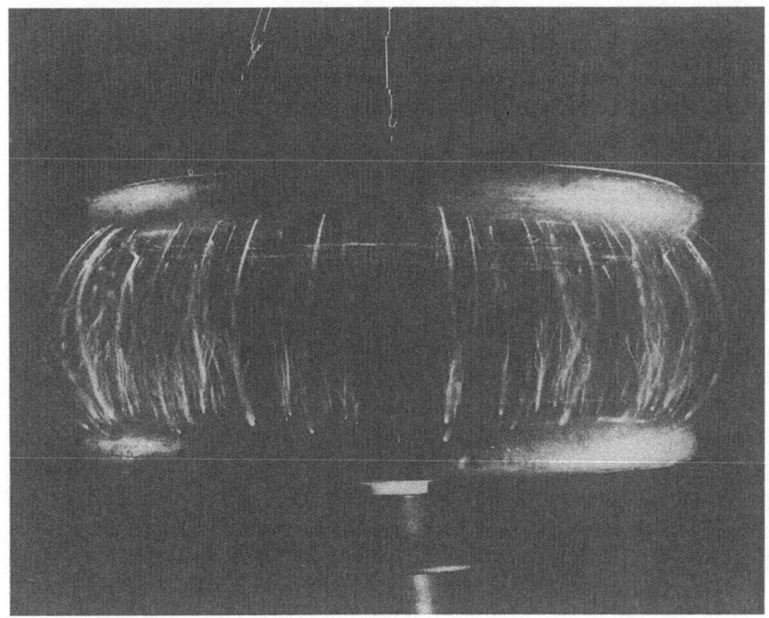

Fig. 3.4-3 Predischarges in a plate-plate gap during stress with a chopped impulse voltage, spacing 200 mm

the electrode regions with maximum field strength [*Marx* 1952]. Fig. 3.4-3 shows this for the example of plate electrodes at a spacing of 200 mm stressed with chopped impulse voltages above 500 kV.

In the primarily practical methods described below, care must be taken to ensure that the measuring cables arranged inside the field zone have only little effect on the electric field. These methods have the definite advantage that they can be applied to manufactured equipment and also take into account the effect of stationary space charges, which appear as a result of the high voltage.

A method for determining the direction of the field strength has been given by *M. Toepler*. It makes use of a small rod-shaped test sample, usually a piece of straw only a few cm long, and is suitable for both direct and alternating voltages. Charges of opposite polarity are induced at the ends of the straw by the field to be measured. This causes a torque which forces the straw, freely suspended at its centre of gravity, to align itself in the direction of the field lines. The position into which the straw test sample is deflected corresponds to the direction of the field strength at that point; parallel projection of the straw onto paper parallel to the plane of rotation enables these directions to be marked. If the straw, suspended by an insulated thread, is displaced in the same rotation plane, and the traces are joined up in accordance with electrostatic field laws, an approximately true reproduction of the field can be constructed.

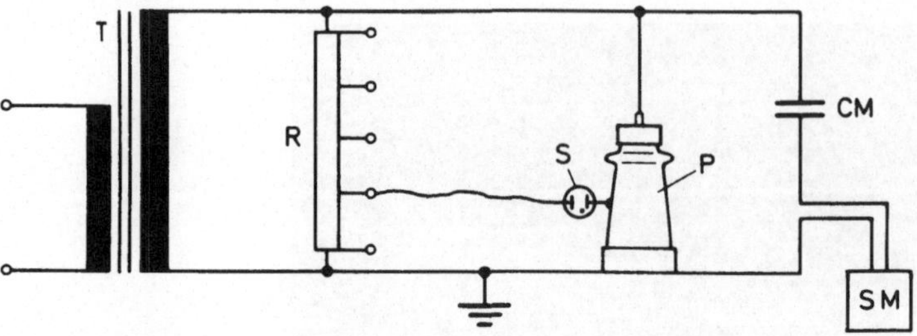

Fig. 3.4-4 Experimental arrangement for measuring the voltage distribution on the surface of high-voltage equipment

T High-voltage transformer
R Resistance according to Fig. 2.4-1a
S Glow lamp of low capacitance
P Test object (porcelain insulating support)
CM Measuring capacitor
SM Peak voltmeter

Measurement of the potential distribution on the surface of insulating bodies can be carried out with the bridge circuit in Fig. 3.4-4. The potential on the surface of the test object P is compared here with the known potential tapped off the divider R. A glow lamp of very low capacitance attached to the surface of the test object P is well suited as zero indicator S. For configurations with rotational symmetry, the capacitive coupling of the glow lamp can be improved by fixing a wire along an equipotential contour on the insulating surface.

3.4.2 Experiment

a) Determination of the Equipotential Lines with Conducting Paper

The circuit of the experimental arrangement is shown in Fig. 3.4-5. The measuring instrument contains a potential divider from which the required partial voltages can be tapped, as well as a zero indicator equipped with an amplifier and an indicating instrument. Probe 2 is connected to the corresponding tap on the divider. The measuring voltage is a few volts and the measuring frequency 50 Hz.

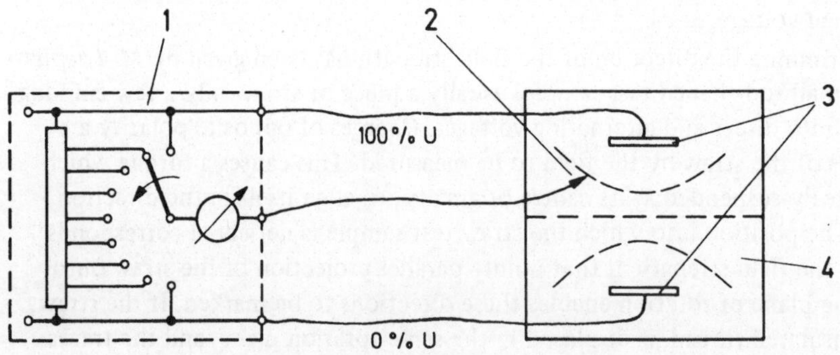

Fig. 3.4-5 Experimental arrangement for measurement of electric fields with conducting paper

1 Bridge for measuring equipotential lines
2 Probe
3 Electrodes
4 Conducting paper

A plane electrode configuration in accordance with special instructions should be reproduced on a board with conducting paper, taking the dielectric constants into account.

Pins should be inserted at the boundary surfaces between different dielectrics. The electrodes should be connected to corresponding potentials on the potential divider of the measuring instrument. The path of the equipotential lines should be plotted and then drawn in.

Fig. 3.4-6 shows the setup during measurement; as an example of a configuration with a two-dimensional field, the base of a current transformer with rated voltage $U_n = 20/\sqrt{3}$ kV is represented. The equipotential lines measured for the same example are shown again in Fig. 3.4-7.

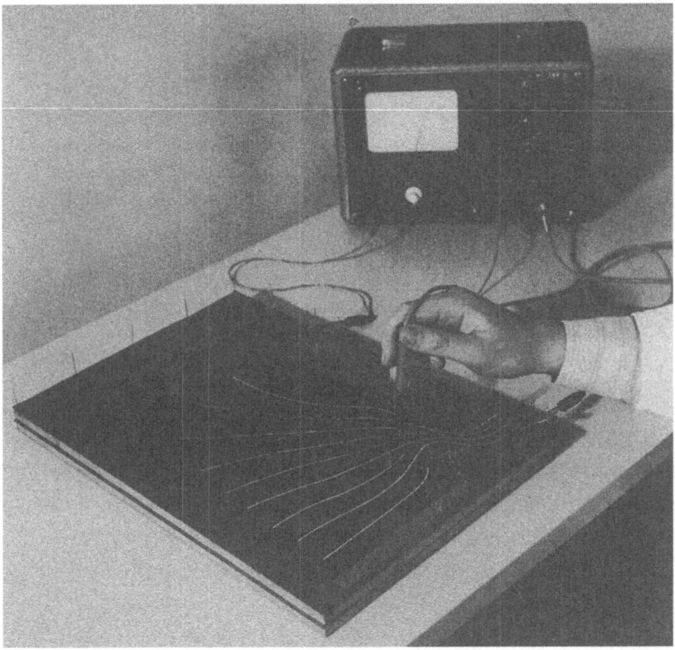

Fig. 3.4-6 Measurement with the equipotential line measuring bridge

b) Measurement of Fields at High Voltage

The circuit shown in Fig. 3.4-4 should be setup. T is a testing transformer for at least 30 kV, CM is a measuring capacitor, SM is a peak voltmeter as in 3.1. As a resistive potential divider R the element shown in Fig. 2.4-1a for example, is suitable. The glow lamp S can be fastened to the surface of the test object with wax. As test object P a 30 kV insulating support is used.

In a trial run, the field plot is obtained using the straw method at an alternating voltage which corresponds approximately to the rated voltage of the test object. In the second

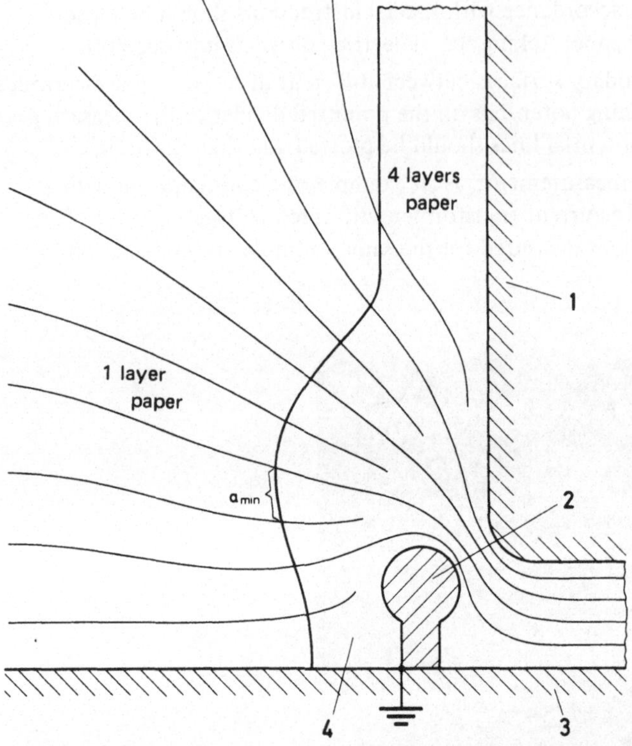

Fig. 3.4-7
Equipotential lines of a current transformer base obtained by simulation with conducting paper
1 High-voltage winding
2 Control electrode
3 Earthed plate
4 Insulation (epoxy resin)

test-series, the glow lamp is fixed to the surface of the test object and connected to R with a clip. During the actual experiment it should be observed that the measuring accuracy depends upon the ratio of the given glow lamp ignition voltage to the total voltage. The total voltage should therefore be at least 20 times the ignition voltage of the glow lamp. The potential of the probe then corresponds to the mean potential selected on R, to within about 5 % of the total voltage.
The traces of the individual positions of the straw shown in Fig. 3.4-8 were obtained in this experiment. In addition, the 25, 50 and 75 % equipotential lines were constructed using some points determined by the glow lamp method and the orthogonality condition for field and equipotential lines.

3.4.3 Evaluation

The points of maximum tangential field strength E_t along the insulator/air boundary should be determined in the equipotential plot obtained with conducting paper according to 3.4.2b. From the example in Fig. 3.4-6, the minimum spacing a_{min} between neighbouring equipotential surfaces is found to be 9 mm. At the rated voltage ($U_n = 20/\sqrt{3}$ kV), the value of E_t regarded as permissible is given by:

$$E_t = \frac{0.1\ U}{a_{min}}\ n = 1.2\ \text{kV/cm}.$$

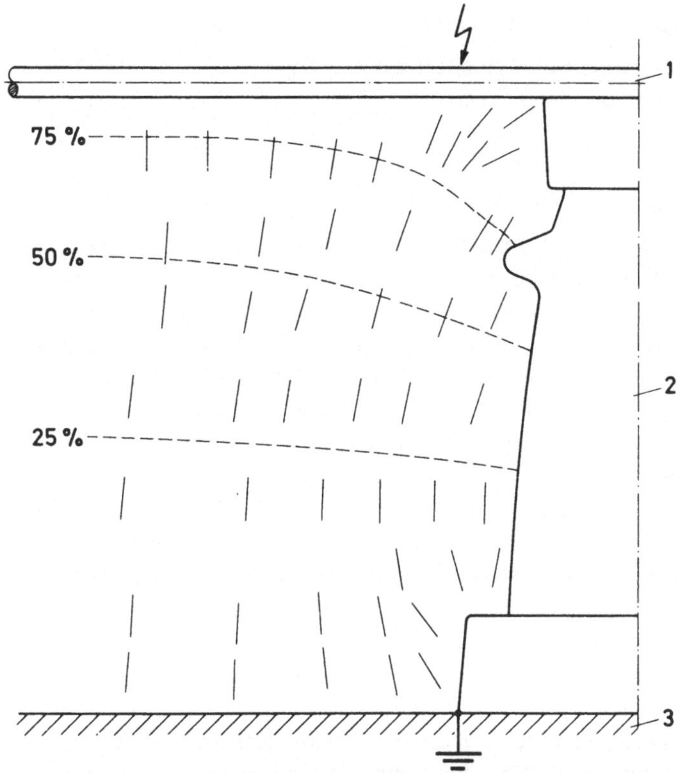

Fig. 3.4-8 Result for measurement of the electric field by the straw and glow lamp methods
1 Reproduction of a busbar 2 Support insulator 3 Earthed base plate

Using the results obtained in 3.4.5 by the straw and glow lamp methods, some equipotential lines should be sketched to a scale of 1 : 1 in the investigated half-plane.

Literature: *Strigel* 1949; *Marx* 1952; *Küpfmüller* 1965; *Potthoff, Widmann* 1965; *Philippow* 1966

3.5 Experiment "Liquid and Solid Insulating Materials"

Insulation arrangements for high voltages usually contain liquid or solid insulating materials whose breakdown strength is many times that of atmospheric air. For the practical application of these materials not only their physical properties but also their technological and constructional features must be taken into account. The topics discussed in this experiment fall under the following headings:

Insulating oil and solid insulating material
Conductivity measurement
Dissipation factor measurement
Fibre-bridge breakdown
Thermal breakdown
Breakdown test

Familiarity with the following sections will be assumed:

1.1 Generation and measurement of high alternating voltages
1.5 Non-destructive high-voltage tests (excluding 1.5.4)

3.5.1 Fundamentals

a) Measurement of the Conductivity of Insulating Oil

The specific conductivity κ of an insulating oil depends strongly on the field strength, temperature and contamination. It is a result of ionic movement and varies in order of magnitude from 10^{-15} to 10^{-13} S/cm for water content ranging from 10 ... 200 ppm[1]). The measurement of κ yields valuable information about the degree of purity of an insulating liquid. The positive and negative ions are produced on dissociation of electrolytic contaminations. For a specific type of ion with charge q_1 and density n_1, the corresponding contribution to the current density at not too high a field strength $\vec{E}$ is given by:

$$\vec{S}_1 = q_1 n_1 \vec{v}_1 = q_1 n_1 b_1 \vec{E}$$

where $\vec{v}_1$ is the velocity and b_1 the mobility of the ions, the latter being constant only when Ohm's law is valid. The corresponding contribution to the conductivity follows:

$$\kappa_1 = q_1 n_1 b_1 .$$

When a certain field strength is established in the dielectric, a compensating mechanism is set in motion to balance the density of the various types of ions and continues until an equilibrium is established between generation, recombination and leakage of ions to the electrodes. Owing to their different mobilities, this compensating mechanism will be realized at different rates for the diverse ions, which is the reason why the resulting conductivity κ is a function of the time after switching on. Fig. 3.5-1 shows the basic characteristic. For measurement of κ it is therefore advisable to wait until these transient mechanisms have passed and begin with the measurement at a certain time, e.g. 1 min, after applying the voltage.

An electrode arrangement which is to be used to measure κ must be fitted with a guard-ring electrode as shown in Fig. 1.5-2. The electric field should be as homogeneous as possible. Apart from plate electrodes, coaxial cylinder electrodes are commonly used. If the measuring voltage applied is U for a homogeneous field of area A and spacing s, κ is derived from the current I:

$$\kappa = \frac{I}{U} \cdot \frac{S}{A} .$$

The currents to be measured are usually of the order of picoamperes. Sensitive moving-coil mirror galvanometers may be used for this purpose. Current measuring devices with electronic amplifiers are easier to handle, however, and much more sensitive.

---

1) ppm = parts per million = 10^{-6}

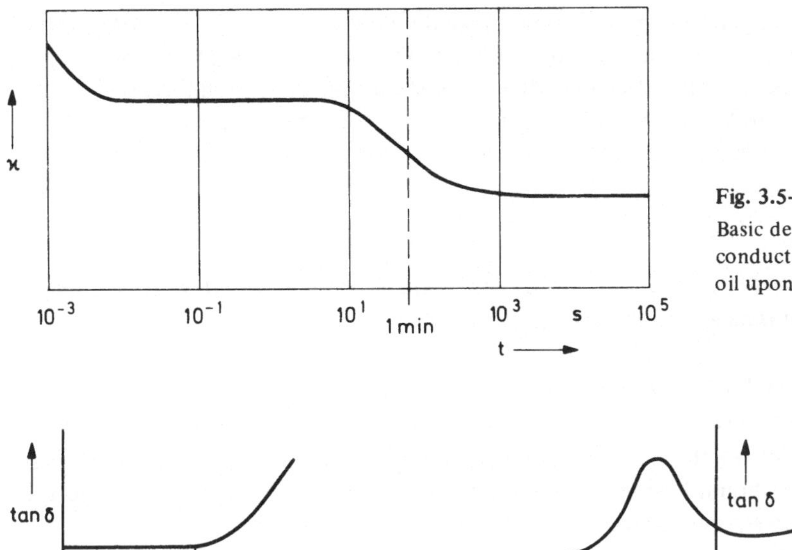

Fig. 3.5-1
Basic dependence of d.c. conductivity of an insulating oil upon the measuring time

Fig. 3.5-2 Basic dependence of the dissipation factor of an insulation upon voltage and temperature
a) $\tan\delta = f(U)$ b) $\tan\delta = f(\vartheta)$

b) Dissipation Factor Measurements of Insulating Oil

Dielectric loss at alternating voltages is due to ionic conduction and polarization losses. The magnitude and nature of these losses, functions of temperature, frequency and voltage, are a measure of the quality of the insulation concerned. They also provide information about the physical mechanisms and permit assessment of the suitability of the insulator for particular applications. Fig. 3.5-2 shows examples of the dependence of the dissipation factor $\tan\delta$ of insulation upon the voltage U and the temperature ϑ, both of which are of great importance in high-voltage technology. From the rise of the function $\tan\delta = f(U)$ at the onset voltage U_e, it can be inferred that partial discharges are initiated either on or within the test sample, causing additional ionic loss. The same shape could however also be a result of field strength dependent variations in the electrolytic conductivity [*Kieback* 1969]. The shape of the function $\tan\delta = f(\vartheta)$ indicates the temperature above which loss due to ionic conduction exceeds that due to polarization.

By definition, the dielectric loss of an insulation with capacitance C at supply frequency ω can be calculated using the dissipation factor:

$$P_{diel} = U^2 \omega C \tan\delta .$$

The loss can be measured in a Schering bridge according to Fig. 1.5-4, which at the same time allows an exact measurement of the test object capacitance, if the capacitance C_N of the loss-free standard capacitor is known.

To determine the dissipation factor of liquid or solid materials, basically the same electrode arrangement used for the measurement of the d.c. conductivity is suitable. The Schering bridge under balanced conditions enables direct reading of the dissipation factor. If the dielectric constant is to be determined, the capacitance C_L of the arrangement in air will also have to be measured in addition to C, so that:

$$\epsilon_r = \frac{C}{C_L} .$$

The relationship $\epsilon_r = f(U)$ or $\epsilon_r = f(\vartheta)$ provides supplementary information concerning the physical mechanisms within the insulating material.

c) Fibre-Bridge Breakdown in Insulating Oil

Every technical liquid insulating material contains macrosopic contaminants in the form of fibrous elements of cellulose, cotton, etc. Particularly when these elements have absorbed moisture from the liquid insulating medium, forces act upon them, moving them to the region of higher field strength as well as aligning them in the direction of $\vec{E}$.

The physical explanation for the alignment of the fibrous elements is the same as that given for the straw method in 3.4.1e.

In this way, fibre-bridges come into existence. A conducting channel is created which can be heated due to the resistance loss to such an extent that the moisture contained in the elements evaporates. The breakdown which then sets in at comparatively lower voltages, can be described as local thermal breakdown at a defect.

This mechanism is of such great technical significance that in electrode arrangements for high voltages pure oil sections have to be avoided. This is achieved by introducing insulating screens perpendicular to the direction of the field strength. In the extreme case consistent application of this principle leads to oil-impregnated paper insulation, which is the most important and very highly stressable dielectric for cables, capacitors and transformers.

d) Thermal Breakdown of Solid Insulating Materials

In solid insulating materials thermal breakdown can be either total, i.e. a consequence of collective overheating of the insulation, or local, i.e. a consequence of overheating at a single defect. It can be explained by the temperature dependence of the dielectric losses; their increase can exceed the rise in the heat being conducted away, P_{ab}, and so can initiate thermal destruction of the dielectric.

Fig. 3.5-3 shows the curves of the power P_{diel} fed in at different voltages and the power P_{ab} which can be drawn from the test object, as functions of the temperature ϑ which is assumed constant throughout the entire dielectric. Thermal breakdown then occurs when no stable point of intersection for the curves of the input and output power exists. Point A represents a stable working condition and point B, on the other hand, is unstable. If the voltage is increased at constant ambient temperature ϑ_u, both points of intersection move closer together until, at $U = U_k$, they coincide in C. This voltage is referred to as the critical voltage; at or above U_k a stable condition is impossible.

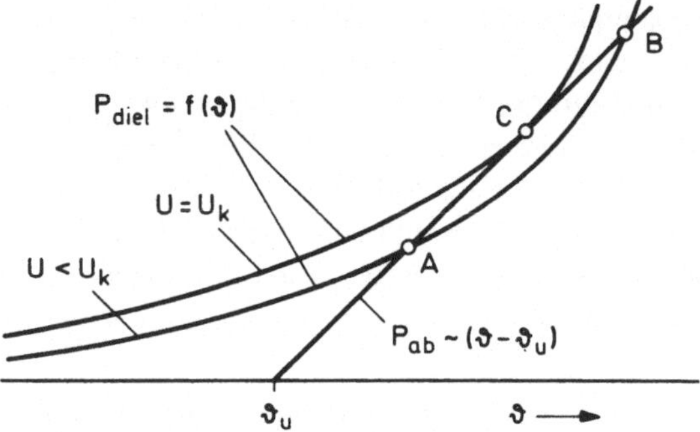

Fig. 3.5-3
Illustration of the thermal breakdown of solid insulating materials

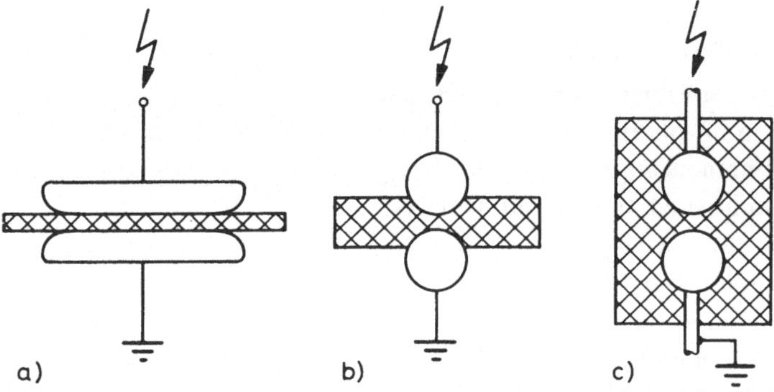

Fig. 3.5-4 Practical test object configurations for breakdown investigations of solid insulating materials
a) Plate electrodes applied to a sample foil
b) Spherical electrodes inserted in a plate sample
c) Spherical electrodes cast in an epoxy resin sample

An increase in $\tan\delta$ at constant voltage shows that U_k has been overstepped at total thermal breakdown. U_k can therefore be experimentally identified without destroying the insulating material. For inhomogeneous field configurations, one should note that the specific dielectric loss P'_{diel} depends upon the square of E:

$$P'_{diel} = E^2 \omega \epsilon_0 \epsilon_r \tan\delta .$$

In regions of maximum field strength, the risk of thermal breakdown is thus particularly high. This can be established in dissipation factor measurements only when the dielectric loss in the endangered area, increasing on account of progressive overheating, is independently measurable, i.e. can be isolated from the overall dielectric loss.

e) Breakdown Strength of Solid Insulating Materials

The experimentally determined values of the breakdown strength of a solid insulating material are, owing to the many possible breakdown mechanisms, strongly dependent upon the electrode configuration in which they have been measured. A particular problem is the fact that the solid insulating material generally has an appreciably higher breakdown strength than the materials in the vicinity of the testing arrangement, so that there is the risk of a flashover. Some simple test object configurations are shown in Fig. 3.5-4.

The arrangement a) where two plate electrodes are applied to a plane solid insulating material, is restricted in its application to very thin insulating foils, a fraction of a mm thick. This is because for larger thicknesses higher voltages would be needed for breakdown, which would lead to gliding discharges at the electrode edges. The breakdown which sets in at these points is more a characteristic of the electrode configuration rather than of the dielectric.

An increase in the onset voltage can be gained by immersing the arrangement in an insulating liquid. The onset of interfering gliding discharges can only be prevented when the product of dielectric constant and breakdown field strength for the immersing medium is greater than that for the solid insulating material to be investigated. The setup can in general be used for breakdown voltages of some 10 kV only.

The onset voltage for gliding discharges at the electrode edges can be raised for plate-shaped solid samples by the insertion of spherical electrodes, either on one or on both sides of the sample. Through additional immersion in a liquid insulating material, e.g. insulating oil, this arrangement as in b) can be used up to about 100 kV.

Plastics have very high breakdown strengths and are even used as homogeneous insulation at working voltages of the order of 100 kV. A suitable test arrangement for epoxy resins is the arrangement c) in which two spherical electrodes are cast into a homogeneous block of insulating material. Additional immersion in a liquid insulating material allows breakdown investigations up to some hundreds of kV to be carried out with this arrangement.

For all arrangements, the advantageous effect of immersion in a liquid insulating material can be improved further, since the breakdown strength of the latter increases under application of higher pressures [*Marx* 1952].

The relevant specifications contain further details for performing the actual breakdown strength measurement.

3.5.2 Experiment

a) Measurement of the d.c. Conductivity of Transformer Oil

In the circuit shown in Fig. 3.5-5a sensitive ammeter, with a built-in d.c. voltage source G[1]), is connected to an oil-filled testing vessel P. A direct voltage U of about 100 V is supplied by the voltage source.

The testing vessel P as in Fig. 3.5-5b consists of a perspex tube 1 with a base of insulating material. A brass bolt 2 with central bore and coaxial socket 3 are fitted through an opening in the base. The brass bolt supports a guard-ring electrode 4. The testing vessel can be closed with a lid 5; the upper electrode 6 is attached to a threaded shaft.

The gap s can be adjusted by turning the threaded shaft in the lid.

1) Picoamperemeter manufactured by Messrs. Knick, Berlin

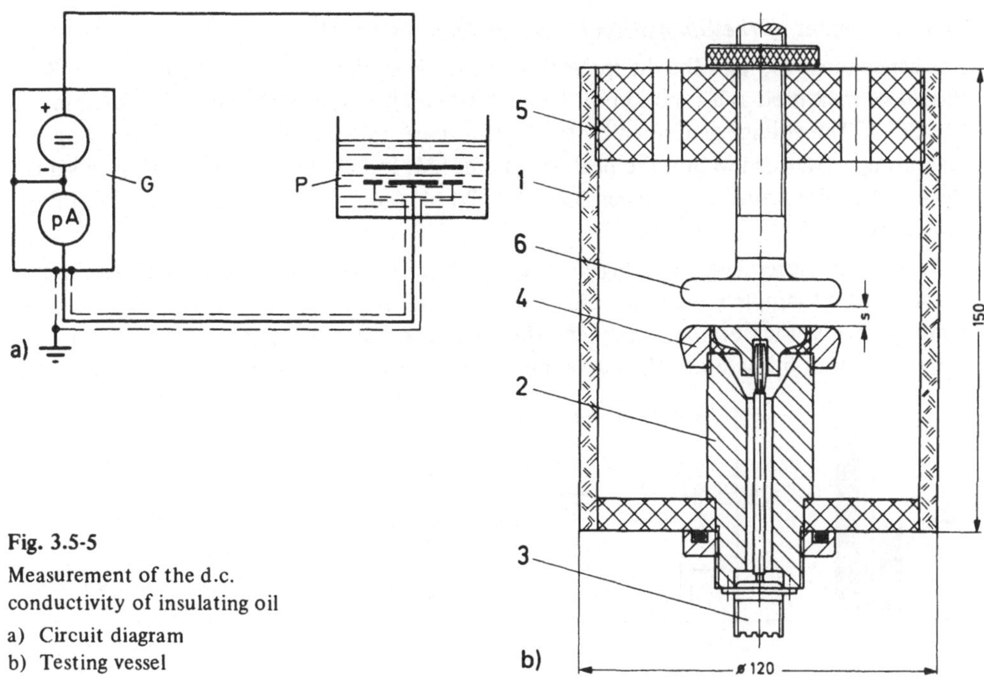

Fig. 3.5-5
Measurement of the d.c. conductivity of insulating oil
a) Circuit diagram
b) Testing vessel

The d.c. conductivity of two different oils should be measured with gap spacings s in the range of 2 ... 5 mm. Fig. 3.5-6 shows the measured curves for the conductivity of two oils as a function of the measuring time; whereas the conductivity of one oil remains constant throughout the measurement, that of the other oil decreases continuously.

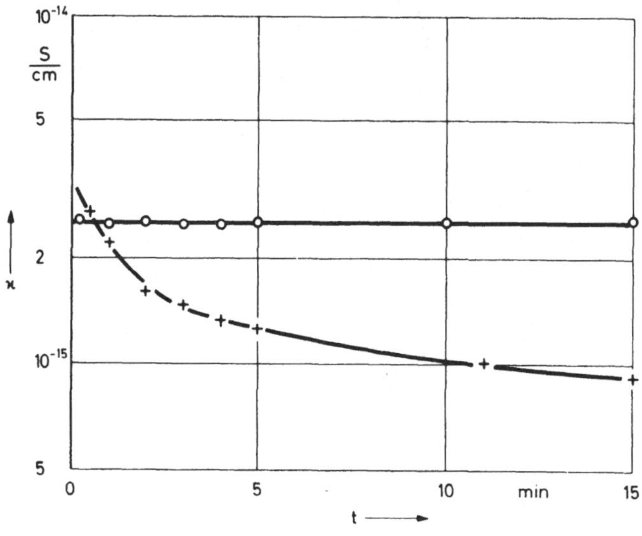

Fig. 3.5-6
Specific conductivity of two transformer oils as a function of measuring time

b) Measurement of the Dissipation Factor of Transformer Oil

The capacitance C_x and the dissipation factor $\tan\delta$ of the arrangement specified in 3.5.2a should be measured as a function of the a.c. test voltage U, using the circuit shown in Fig. 3.5-7. The voltage generated by the high-voltage transformer T is measured using the measuring capacitor CM and the peak voltmeter SM. In parallel with the testing vessel is the standard capacitor with capacitance $C_2 = 28$ pF (compressed gas capacitor as in Fig. 2.4-3).

A series of measurements at voltages up to 35 kV and spacing s = 5 mm on a repeatedly pre-stressed oil sample yielded the curve $\tan\delta = f(U)$ shown in Fig. 3.5-8. The curve, a result of regular measurements at increasing and decreasing testing voltage, shows distinct hysteresis. Both branches of the curve, however, increase continuously with the voltage.

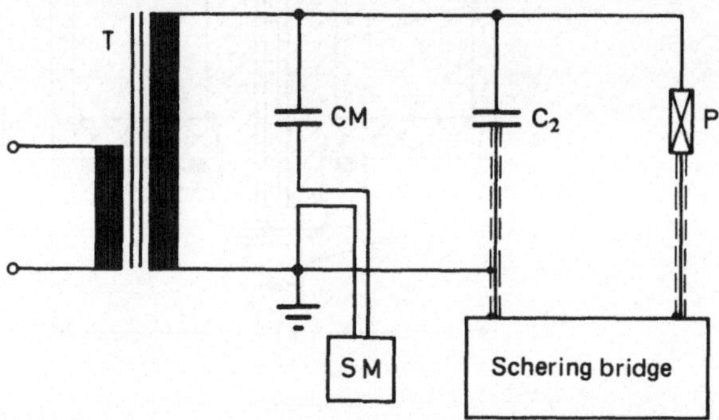

Fig. 3.5-7 Circuit for measuring the capacitance and dissipation factor of a test object with the Schering bridge

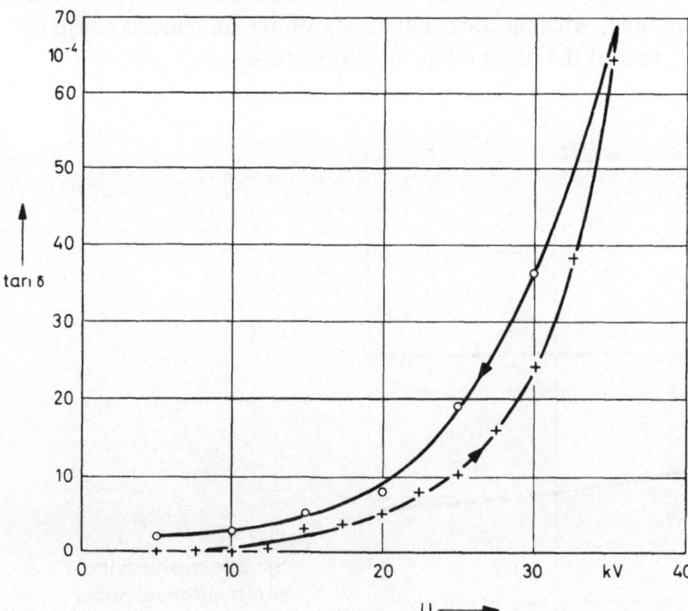

Fig. 3.5-8 Dependence of the dissipation factor of a transformer oil upon the test voltage

From about 27 kV onwards, the influence of the measuring time on the dissipation factor becomes noticeable. During each measurement the applied voltage was held constant for 2 minutes. The values measured at the beginning of each such interval were lower than at the end; the mean values were plotted in the diagram.

c) Fibre-Bridge Breakdown in Insulating Oil

In the setup shown in Fig. 3.5-5b, the upper electrode is replaced by a sphere, e.g. of 20 mm diameter, and the spacing set to a few cm. Some slightly moistened black pieces of cotton 5 mm long are contained in the oil. A voltage of about 10 kV applied between the sphere and the plate within seconds results in the alignment of the threads in the direction of the field: a fibre-bridge is established, which can either initiate or accelerate a breakdown. The two photographs of the model experiment shown in Fig. 3.5-9 indicate clearly the extent to which oil gaps in high-voltage apparatus, which are not subdivided, are exposed to risk by dissociation products and other solid particles.

d) Breakdown of Hardboard Plates

The 1-minute withstand voltage of 1 mm thick plates of a hardboard sample should be determined as follows. The test circuit is the same as in 3.5.2c.

The breakdown voltage should be determined approximately in two preliminary trial runs with a rate of voltage rise of 2 ... 3 kV/s. The resulting mean value, as breakdown voltage

a)

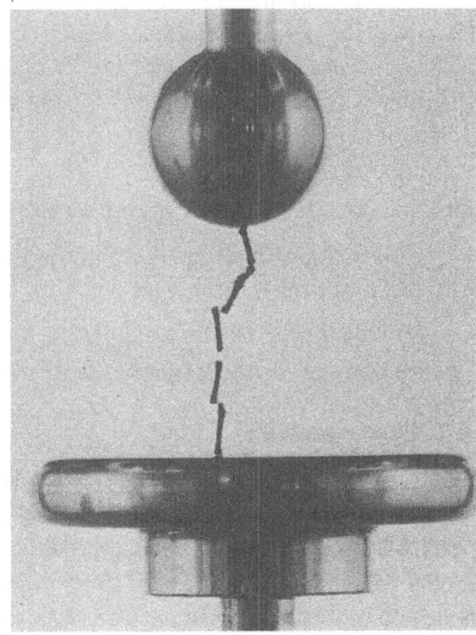

b)

Fig. 3.5-9 Model experiment showing fibre-bridge formation in insulating oils

a) Fibres before switching the voltage on

b) Fibre-bridge 1 minute after switching the voltage on

U_{dm}, will be taken as the basis of future experiments. In the first minute of stress the voltage 0.4 U_{dm} should be applied. Then the voltage should be increased by 0.08 U_{dm}, held for 1 minute, and so on, until breakdown occurs. The voltage at which the insulation was just on the verge of breakdown is the 1 minute withstand voltage. The 5 minute withstand voltage should be determined in a similar way; it is appreciably lower as a rule.

3.5.3 Evaluation

The time-dependent characteristic of the d.c. conductivity of transformer oils, measured as in 3.5.2a, should be graphically represented.
The dissipation factor and the capacitance of the oil-filled testing arrangement, measured as in 3.5.2b, should also be shown graphically as a function of the voltage.
The formation of fibre-bridges should be noted in the testing arrangement as in 3.5.2c.
The 1 minute and the 5 minute withstand voltages of 3 plates of a 1 mm thick hardboard sample should be determined according to the method given in 3.5.2d.
The ratio of the two withstand voltages should be calculated from the mean values and discussed.

Literature: *Whitehead* 1951; *Böning* 1955; *Imhof* 1957; *Lesch* 1959; *Roth* 1959; *Anderson* 1964; *Potthoff, Widmann* 1965

3.6 Experiment "Partial Discharges"

In insulation systems with strongly inhomogeneous field configurations or inhomogeneous dielectrics, the breakdown field strength can be locally exceeded without complete breakdown occurring within a short time. Under these conditions of incomplete breakdown the insulation between the electrodes is only partially bridged by discharges. These partial discharges (PD) have considerable practical significance, particularly for the case of stress by alternating voltages.
The topics covered in this experiment fall under the following headings:

External partial discharges (Corona)
Internal partial discharges
Gliding discharges

It is assumed the reader is familiar with section 1.5: Non-destructive high-voltage tests.

3.6.1 Fundamentals

In a strongly inhomogeneous field external partial discharges occur at electrodes of small radius of curvature when a definite voltage is exceeded. These are referred to as corona discharges and, depending upon the voltage amplitude, they result in a larger or smaller number of charge pulses of very short duration. It is these discharges which are the source of the economically significant corona losses in high-voltage overhead lines; moreover, the electromagnetic waves generated by the charge pulses can also cause radio interference.
In high-voltage equipment partial discharges can also occur at a distance from the electrode surfaces, particularly in gas inclusions in solid or liquid insulating materials (cavities, gas

bubbles). Hence there is the risk of damage to the dielectric as a result of these internal partial discharges during continuous stress, due to breakdown channels developing from such partial discharge sites and because of additional heating. Partial discharges which develop at the interface of two dielectrics in different states of aggregation are known as gliding discharges. Especially when the interface under stress is in close capacitive coupling with one of the electrodes, high energy discharges take place which, even at moderate voltages, can bridge large insulation lengths and so damage the insulating materials.

a) Partial Discharges at a Needle Electrode in Air

The most important physical phenomena of external PD at alternating voltages can be observed particularly well for the example of a needle-plate electrode configuration in air. Fig. 3.6-1 shows a suitable configuration.

As curve 1 Fig. 3.6-2 schematically shows, when the applied voltage is increased pulses first appear at the peak of the negative half-period, the amplitude, form and periodic spacing of which are practically constant. These are the so-called "Trichel Pulses", also observed for negative direct voltages, on the evidence of which *G. W. Trichel* in 1938 demonstrated the pulse-type character of corona discharges.

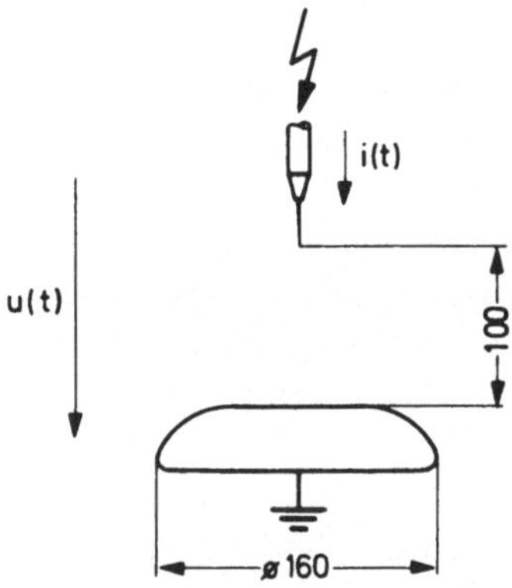

Fig. 3.6-1 Needle-plate gap

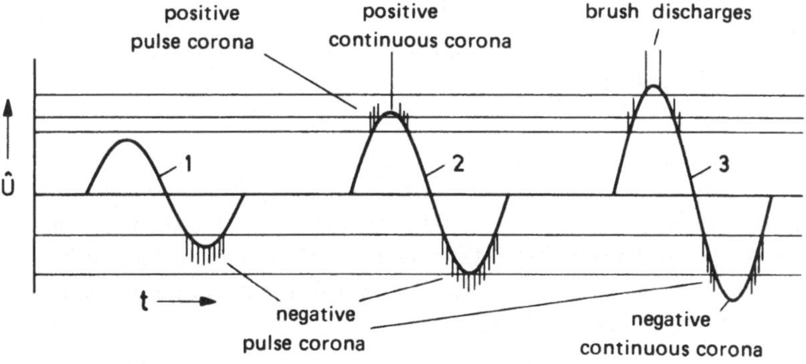

Fig. 3.6-2 Type and phase relation of partial discharges

The pulse length is a few tens of ns and their frequency can be up to 10^5 Hz. If the voltage is increased further, pulses also appear at the peak of the positive half-period, yet these are irregular (curve 2).

For both polarities, with increasing voltage in the peak region pulse-less partial discharges may also occur, referred to as "continuous corona" (curves 2 and 3); this is the reason for the comparatively low radio interference found in some cases, despite extensive corona losses. The final typical discharge mode prior to complete breakdown is intense brush discharge in the positive peak (curve 3).

The pulse-type character of the pre-discharges may be explained using the Trichel-pulse example. The electron avalanches produced at the negative point electrode travel in the direction of the plate. Their velocity is strongly reduced owing to the rapidly decreasing field strength (see Fig. 3.2-2), and by attachment of electrons to the gas molecules, negative ions are formed. The space charge so produced reduces the field strength at the cathode tip, thus preventing further formation of electron avalanches. A new electron avalanche can commence from the cathode only after removal of the space charge by recombination and diffusion. The pulse-type discharges occur in the region of the test voltage peak.

b) Corona Discharges in a Coaxial Cylindrical Field

Corona performance of overhead line conductors is of great significance to the technical properties and economics of a high-voltage line. Corona measurements can be carried out in the laboratory, if the conductor arrangement to be studied is chosen to be the inner electrode of an assembly of coaxial cylinders. In this "corona-cage" the field configuration near the conductor differs very little from that of the actual transmission line, since one may safely assume that the conductor spacing in the latter is very large compared with the conductor radius and therefore the field in the vicinity of the conductor similarly possesses cylindrical symmetry.

Fig. 3.6-3 shows an arrangement which can be used for a.c. experiments up to about 80 kV. The conductor 1 to be studied is stretched along the axis of the outer cylinder 2 and connected to the alternating voltage $u(t)$. The current i in the earth lead of the insulated outer cylinder is measured. It may be assumed that this current approximately corresponds to that flowing from the high-voltage conductor. For exact measurements the corona-cage should be put into effect as a guard-ring arrangement.

The current i comprises the displacement current and the corona current, the capacitance is assumed to be constant [*Sirotinski* 1955]:

$$i = C\frac{du}{dt} + i_k .$$

The corona current i_k increases rapidly with the instantaneous value of the voltage, once the onset voltage U_e is exceeded. It results from the migration of the ions formed by the discharge in the previous or in the same half-period. Fig. 3.6-4 shows the current characteristic expected for this considerably simplified treatment.

i_k is an effective current and corresponds to the corona losses. These are caused by the power required to maintain collision ionisation and by the conductor current, represented

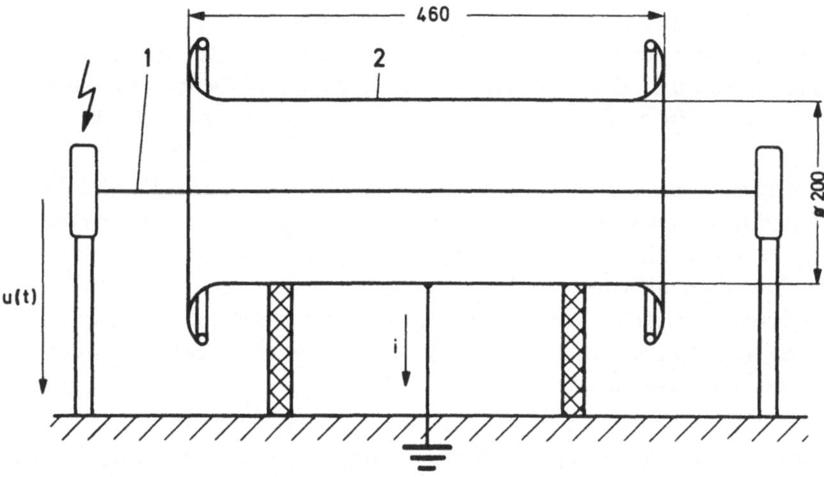

Fig. 3.6-3 Corona-cage. 1 Inner conductor 2 Outer cylinder

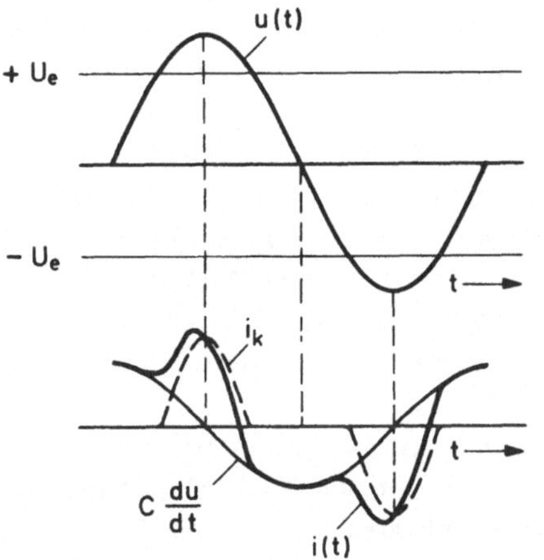

Fig. 3.6-4
Voltage and current characteristics in the corona-cage

by the movement of charge carriers. Corona losses in overhead lines are strongly dependent upon weather conditions and can deviate from the annual mean value up to one order of magnitude above or below. The charge carriers emerging from the collision ionisation region, by attachment to neutral gas molecules, form large ions which are accelerated away from the corona electrode; an "electric wind" is produced. This phenomenon has acquired great practical significance in the electrostatic purification of gases.

c) Partial Discharge Measurement in High-Voltage Insulation Systems

Partial discharges on or in a test object have become an important research topic of high-voltage technology since they can be an indication of manufacturing defects in electrical equipment or the cause of ageing of an insulation. Details for the conduction of PD measurements in connection with insulation tests at alternating voltages are given in IEC Publ. 270. For radio interference tests other aspects apply.

The most important PD measurements of high-voltage equipment aim to determine the onset voltage U_e and the extinction voltage U_a. In practical arrangements however, the onset and extinction of partial discharges are usually not very distinct phenomena. These measurements therefore require an accompanying declaration of the sensitivity of the methods used.

If a large number of PD sites are present in an insulation system, a noticeable increase of the losses in the dielectric occurs when the onset voltage range is exceeded. The magnitude of this increase is a measure of the intensity of the partial discharges, so long as the basic dielectric losses are low or remain constant. The Schering bridge is therefore also used for the measurement of corona losses in overhead lines or for the measurement of ionisation losses in cables, when these contain numerous distributed defects as a consequence of the manufacturing process (non-draining compound-filled cables).

To record and assess PD in technical insulation systems with individual defects, more sensitive measuring methods should be applied. For this purpose instruments are used which amplify the high frequency electrical disturbances initiated by the partial discharges, and evaluate these in various ways. The measuring instrument is coupled as a rule by an ohmic resistance R, connected either as in Fig. 1.5-10 to the earth lead of the test object or to that of a coupling capacitor.

In a real test object the voltage at R, as a consequence of the partial discharges, consists of an irregular series of pulses of very different amplitudes; their length is dependent upon the characteristics of the circuit and can be some tens of ns. The objective of the PD measuring technique is to register this statistical quantity and evaluate it in view of the evidence desired. Various evaluation methods have been recommended for this purpose: e.g. a selective or wide-band measurement of the peak value or arithmetic mean value of the voltage pulses, the charge or repetition rate of the pulses and derived quantities such as PD power. By calibrating with pulse generators one aims to estimate the effect of the characteristics of the total setup upon the measured result.

One method, preferentially applied in testing bays, makes use of selective interference voltage measuring instruments for weighting the measuring quantity at R, developed for measurement of radio interference voltages. These instruments are constructed as shown in the block circuit diagram of Fig. 3.6-5; weighting follows by making use of the physiological effect of the disturbance on the human ear.

d) Gliding Discharges

One may always expect gliding discharges when high tangential field strengths appear at interfaces. For some insulating assemblies in high-voltage technology, a flashover can be induced for this reason. Two typical examples of this are shown in Fig. 3.6-6. The basic shape of the equipotential lines can be illustrated by the partial capacitances pertaining to a virtual intermediate electrode A. Since the surface capacitance C_0 is much greater than C_1, almost the entire voltage appears at C_1.

When the onset voltage is exceeded, partial discharges occur which develop with increasing voltage from corona to brush discharges along the surface. The intensity of these gliding discharges and their onset voltage depend upon the magnitude of the surface capacitance C_0. The larger it is, the larger too, for time-varying voltages, is the discharge current which

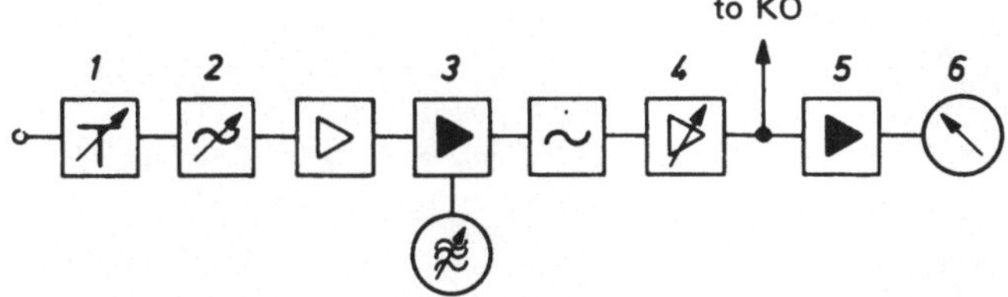

Fig. 3.6-5 Block circuit diagram of a selective interference voltage measuring instrument
1 Input with variable damping element (calibration divider)
2 Tunable input circuit (filter)
3 Amplifier, oscillator, mixing stage
4 Intermediate frequency amplifier with variable amplification
5 Weighting element
6 Indicating arrangement

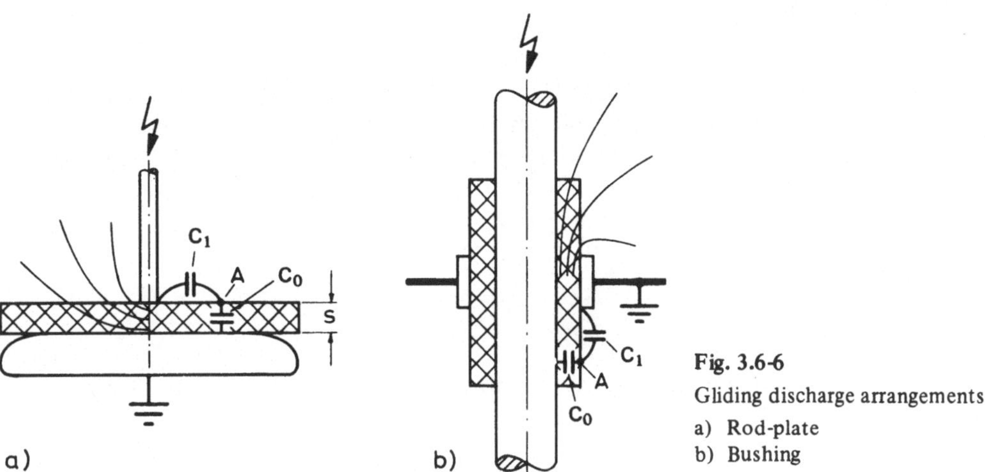

Fig. 3.6-6
Gliding discharge arrangements
a) Rod-plate
b) Bushing

flows from the tip of the brush discharge through the insulator as a displacement current. This leads to extension of the high-voltage potential on the surface, without an appreciable reduction occurring in the field strength at the tip of the discharge. Further growth of the discharge is thus favoured.

For direct voltages gliding discharges scarcely occur, if at all, owing to the absence of displacement currents. The decisive role is played here by the surface conductivity.

For impulse voltages, the rapid voltage variations lead to particularly large displacement currents, which is why the gliding discharges in this case have a very high energy. From the shape and range of the gliding discharges, it is possible to deduce the polarity and the amplitude of an impulse voltage; this fact is made use of for measuring purposes in klydonographs. Here, in an electrode arrangement similar to Fig. 3.6-6a with a point high-voltage electrode, the upper surface of the insulating plate is coated with an active photochemical or dust-like layer. Lichtenberg figures, two examples of which are reproduced in Fig. 3.6-7, are obtained in this way. These show clearly the distinct polarity dependence of the gliding discharge mechanism [*Marx* 1952; *Nasser* 1971].

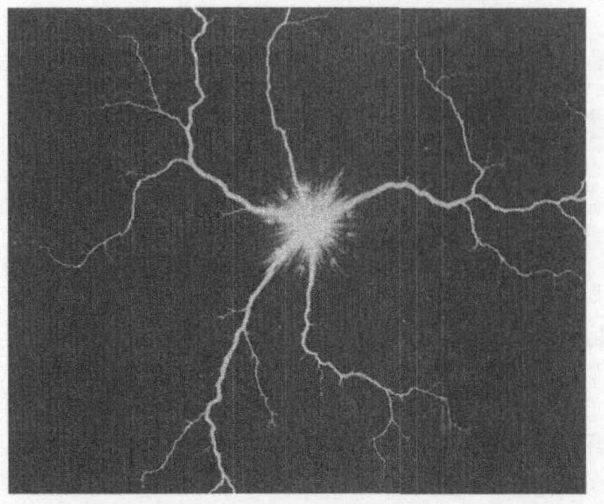
a)

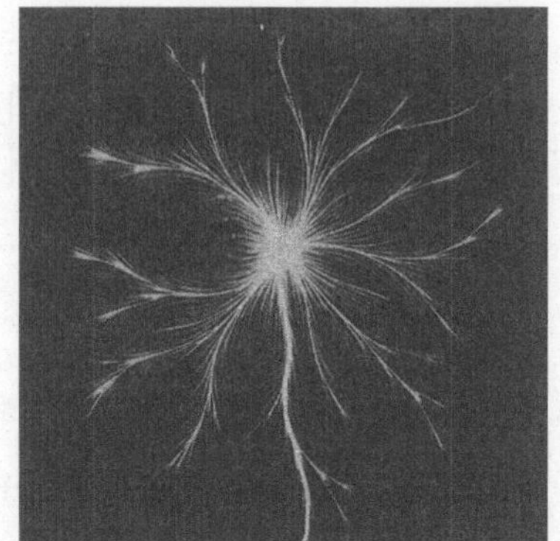
b)

Fig. 3.6-7 Lichtenberg figures (after Marx 1952)
a) Positive point b) Negative point

The determination of the onset voltage U_e for the different discharge phases in a gliding discharge arrangement at alternating voltages, is of particular significance to the design of an insulation system. As shown by *M. Toepler* in 1921, U_e decreases with increasing magnitude of the surface capacitance. For the plane configuration with sharp-edged high-voltage electrode, as in Fig. 3.6-6a, the following empirical relationship is valid, with U_e in kV and s in cm [*Kappeler* 1949; *Böning* 1955]:

$$U_e = K\left(\frac{s}{\epsilon_r}\right)^{0.45}.$$

The values of K depend on material and are different for each discharge phase. They are, approximately:

Corona onset: K = 8 for metal edge in air
K = 12 for graphite edge in air
K = 30 for metal or graphite edge in oil
Brush discharge onset: K = 80 for metal or graphite edge in air or oil

Overstepping the brush discharge onset voltage often leads to permanent damage of the insulating surface within a very short time.

3.6.2 Experiment

The following circuit elements are used repeatedly in this experiment:

- T Testing transformer 220 V/100 kV, 5 kVA
- SM Peak voltmeter (see 3.1)
- CM Measuring capacitor 100 pF, 100 kV
- KO Cathode-ray oscilloscope
- STM Interference voltage measuring device STTM 3840 a (manufactured by Siemens), measuring frequency 30 kHz to 3 MHz (set to 1.9 MHz), bandwidth 9 kHz
- AV Coupling four-pole STAV 3856 (60 Ω) (manufactured by Siemens)

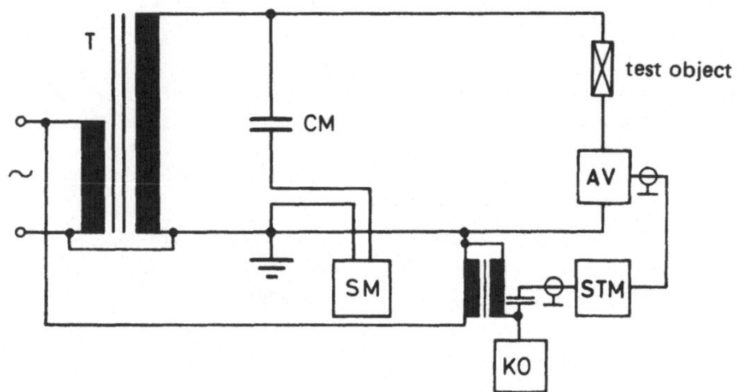

Fig. 3.6-8
Test circuit for partial discharge measurements

All measurements are performed with alternating voltages in the circuit shown in Fig. 3.6-8, but with different test objects.

a) Partial Discharges at a Needle Electrode in Air

A needle-plate gap, as in Fig. 3.6-1, with spacing s = 100 mm is incorporated as test object. The high-voltage electrode consists of a rod with a conical tip, into which a sewing needle has been inserted. The interference voltage measuring device should be connected to the four-pole AV in the earth lead of the test object. The various interference voltage pulses could be taken from the intermediate frequency amplifier of the device constructed according to Fig. 3.6-5. By time dilation of the signal, a convenient oscillographic indication of the onset of pulses is possible. Corresponding to Fig. 3.6-8, the pulses are capacitively superimposed on an alternating voltage in phase with the test voltage, so that their phase relation with respect to the test voltage can be shown on the KO. A tolerably good recording of the pulse shape itself calls for measuring equipment with bandwidths of at least 100 MHz.

The discharge patterns at the needle for each voltage range are observed by varying the test voltage, and compared with the schematic representation of Fig. 3.6-2.

b) Measurements in the Corona-Cage

The corona-cage as in Fig. 3.6-3 should now be connected as the test object. A bare copper wire of diameter d = 0.4 mm is inserted as the inner electrode. In a first series of measurements the PD interference voltage U_{PD} is measured as a function of the test voltage. At the same time the phenomena of incomplete breakdown should be observed up to the onset of complete breakdown.

In a second series of measurements the coupling four-pole is replaced by a screened measuring resistor, to which a capacitor and a surge diverter are connected in parallel for overvoltage protection. The time constant RC should be about 100 μs.

At increasing test voltage the time-dependent curve of the cage current i should be observed for up to about 80 % of the breakdown voltage, and recorded at a voltage U which produces a particularly distinct current curve. Fig. 3.6-9 shows an oscillogram of the cage current at U = 40 kV. The curve confirms the ideas described in paragraph 3.6.1b.

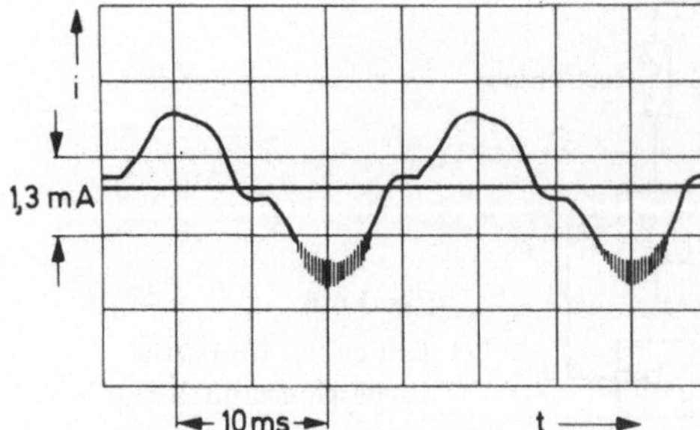

Fig. 3.6-9
Oscillogram of the cage current
U = 40 kV, diameter of the inner conductor d = 0.4 mm

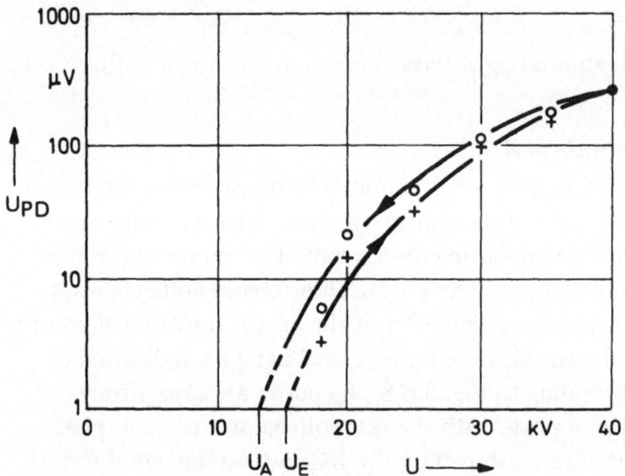

Fig. 3.6-10
Curve of the interference voltage of a 20 kV current transformer

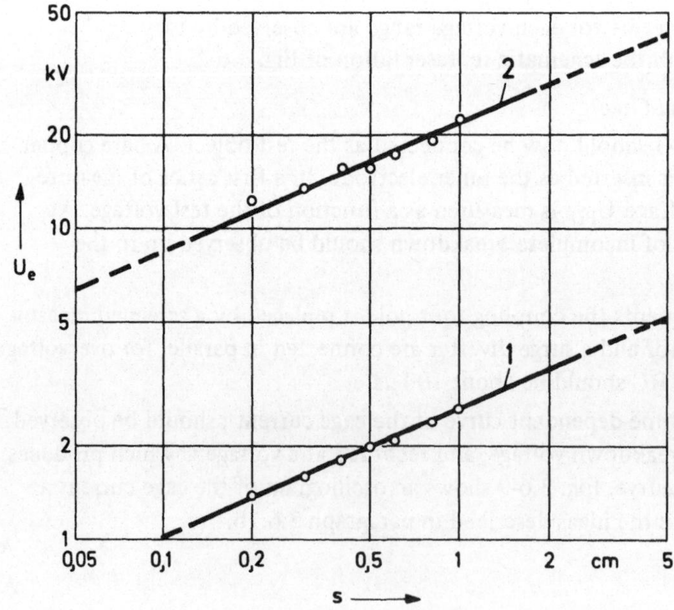

Fig. 3.6-11
Onset voltage of a gliding discharge arrangement as in Fig. 3.6-6a

1 Corona onset
2 Brush discharge onset

c) Partial Discharge Measurements on a High-Voltage Apparatus

A 20 kV current transformer should be connected as test object; its high-voltage terminals should be fitted with screening electrodes if necessary, to avoid external partial discharges. The earthing connection of the test object is again made via the coupling four-pole; the interference voltage measuring device is connected.

The PD interference voltage U_{PD} should be measured for up to 90 % of the test voltage stated on the rating plate of the test object. The voltage should then be reduced at about the same rate and in doing so U_{PD} determined again. U_e and U_a should be measured. The curves shown in Fig. 3.6-10 were obtained for this experiment.

d) Measurement of the Onset Voltages of Gliding Discharges

The test object should be arranged according to Fig. 3.6-6a with glass plates in air as the dielectric. The relationship $U_e = f(s)$ should be measured for various plate thicknesses $s = 2, 3, 4, 5, 6, 8$ and 10 mm. The onset of gliding discharges in Fig. 3.6-11 should be determined with the STM and that of the brush discharges visually.

By logarithmic graduation of the coordinates, the measured points can be represented quite well by straight lines. This corresponds to the relationship given in section 3.6.1d:

$$U_e \sim s^{\text{const.}}$$

With $\epsilon_r \approx 10$ for line 1 one obtains $K \approx 8$, and for line 2: $K \approx 70$. Deviations to higher values for the corona onset voltage can occur for insulating materials with high surface resistance, such as glass; this may be explained by the formation of surface charges.

3.6.3 Evaluation

Using the measurements from section 3.6.2b the breakdown strength E_d of the wire used as the inner conductor of the cage should be calculated.

From the time variance of the current i, recorded as in section 3.6.2b, the approximate separation of the two components according to section 3.6.1b should be attempted.

From the results obtained in section 3.6.2c, the relation $U_{TE} = f(U)$ should be plotted in a diagram for increasing and decreasing test voltages of the current transformer.

The relation $U_e = f(s)$ measured in section 3.6.2d for the onset of corona and brush discharges should be plotted in double-logarithmic representation.

Literature: *Gänger* 1953; *Sirotinski* 1955; *Roth* 1959; *Nasser* 1971; *Schwab* 1972

3.7 Experiment "Breakdown of Gases"

The analysis of the breakdown of gases is also important for understanding the breakdown mechanisms in liquid and solid insulating materials. Gas discharges always occur after the onset of breakdown in any type of dielectric. Gases have a wide range of application as insulating media, especially as atmospheric air. The topics covered by this experiment fall under the following headings:

Townsend mechanism
Streamer mechanism
Insulating gases

It is assumed the reader has some basic knowledge of the mechanisms of electrical breakdown of gases, as well as acquaintance with section 1.1: Generation and measurement of alternating voltages.

3.7.1 Fundamentals

a) Townsend Mechanism

The breakdown of gases at low pressures and small spacings can be described by the Townsend mechanism. Thereby, electrons of external origin accelerated by the field can form new charge carriers by collision ionisation, provided their kinetic energy exceeds the ionisation potential of the gas molecules concerned. An electron avalanche is built up which travels from the cathode to the anode. If, as a consequence of the avalanche, a sufficient number of new ions are formed near the cathode, complete breakdown finally takes place.

It can be shown that for this kind of discharge formation the static breakdown voltage U_d of a homogeneous field at constant temperature depends only upon the product of pressure p and spacing s.

The ionisation coefficient of the electrons α and its dependence upon the field strength E can be described by the formula:

$$\frac{\alpha}{p} = A e^{-B\frac{p}{E}}$$

where A and B are empirical constants. For the Townsend mechanism in a homogeneous field, the following breakdown condition is valid:

$$\alpha s = k = \text{const.}$$

When this equation is satisfied, $E = E_d = U_d/s$. Substituting and solving for U_d, one obtains the Paschen law:

$$U_d = B \frac{ps}{\ln (A/k\ ps)} = U_d(ps).$$

Whether or not the conditions of this law are satisfied can be taken as evidence for or against a discharge occurring by the Townsend mechanism.

b) Streamer Mechanism

At higher pressures and larger spacings discharge in gases takes place by the streamer mechanism according to Raether, Loeb and Meek. It is characterized by the fact that photon emission at the tip of an electron avalanche induces new avalanches and initiates the growth of a streamer at a very fast, abrupt rate, compared to the growth of the primary avalanche.

The onset of photo-ionisation, very effective for the growth of the discharge, should be expected when the multiplication factor of the avalanche, $e^{\alpha x}$, has reached a critical value of about $e^{20} \approx 5 \cdot 10^8$.

The transition of a discharge from Townsend growth to streamer growth can, for a given spacing, be promoted by several parameters:

The larger the product ps, the smaller is the probability that an individual avalanche can traverse the discharge space before critical multiplication is reached. For overvoltages up to about 5 % above the static breakdown value of U_d, a discharge in air by the Townsend mechanism may be expected only for values of

$$ps \leqslant 10 \text{ bar mm}.$$

At higher values, breakdown occurs by the streamer mechanism.

For steep impulse voltages, high field strengths can appear locally which lie well above the static value of E_d, depending upon the impulse voltage-time curve of the arrangement. α increases strongly with E and consequently critical multiplication will be reached even in a short avalanche length.

The ionisation probability of photon radiation is approximately proportional to the density of the gas. Therefore the greater the product of molecular weight M and pressure p, the sooner critical multiplication of the avalanche and with that its change-over to streamer growth takes place.

High field strengths already prevail in strongly inhomogeneous fields near electrodes with strong curvature before the ignition of a self-sustained discharge. Thus it can be shown that for spherical and cylindrical electrodes, E_d increases rapidly with decreasing radius of curvature r. It follows that an avalanche, once started, easily reaches critical multiplication.

The ionising radiation, which causes the change-over to streamer growth, considerably promotes the formation of a streamer discharge.

c) Types of Gas Discharges

The resistance of a gas-filled gap collapses to low values once the voltage for complete breakdown is reached. The type of gas discharge which then occurs and its duration depend upon the yield of the energizing current source.

When currents of the order of 1 A or more flow in the discharge path, one may expect arc discharges. In this case a well-conducting plasma column develops as a result of thermo-ionisation, the arc voltage of which decreases with increasing current.

If the current flowing after breakdown lies in the mA-range, one may expect glow discharges, particularly at low gas pressures (e.g. 100 mbar). For this type of discharge the charge carriers are formed by secondary emission at the cathode. A general statement concerning the current dependence of the arc voltage cannot be made.

The discontinuous transition to a discharge with higher current is referred to as spark discharge. In breakdown processes this is usually the transition to the arc discharge, which only lasts for a short while during voltage tests however. On the other hand, in power supply networks the extinction of a once established arc is usually only effected after switch-off.

d) Gases of High Breakdown Strength

Dry air or nitrogen are cheap insulating materials of high electrical strength, particularly at high pressures, which therefore find extensive technical application. One may mention metal clad switch gear, compressed-gas capacitors or physics apparatus as examples. In all these cases, however, the mechanical stress to which the large containers are subjected calls for considerable constructional measures.

For homogeneous or only slightly inhomogeneous electrode configurations in air or nitrogen in the usual range of gap spacings of the order of centimetres, an increase in pressure beyond about 10 bar results in progressive deviation from the Paschen law. The breakdown voltage U_d no longer increases in proportion with p, as shown in Fig. 3.7-1a. The reason for this probably rests with the associated ideas mentioned under 3.7.1b. For extremely inhomogeneous configurations a pressure increase could even lead to reduction of U_d. In this case promotion of the discharge growth by photo-emission predominates over the obstruction of collision ionisation owing to the increased pressure. Fig. 3.7-1b shows a schematic representation of the possible curve.

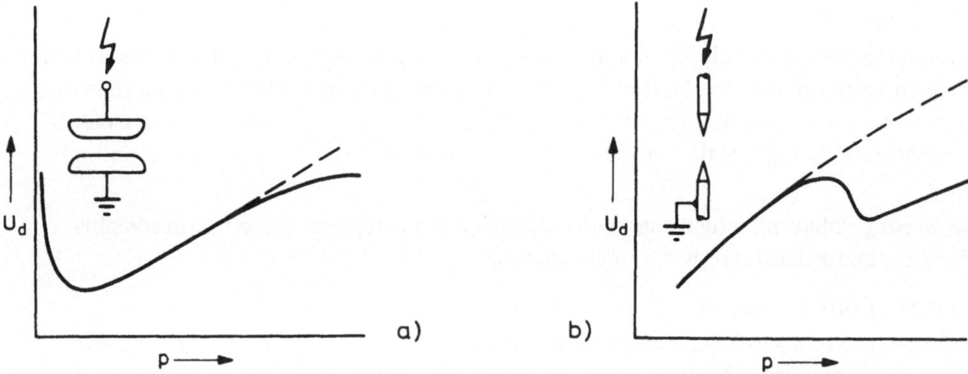

Fig. 3.7-1 Breakdown voltage of a gas as a function of pressure
---- Curve according to the Paschen law. a) Homogeneous field b) Inhomogeneous field

The excellent properties of sulphurhexafluoride (SF_6) for insulations and for arc-quenching have been known for a long time. Nevertheless widespread application of this highly electronegative gas has only been in progress since about 1960. It is used for the insulation of high-voltage switch gear, high-power cables, transformers and large-size physics equipment, as well as for arc-quenching in power circuit breakers.

SF_6 has a molecular weight of 146 and is composed of 22 % by weight of sulphur and 78 % of fluorine. It is built up in such a way that the sulphur atom is at the centre of a regular octahedron, with fluorine atoms at each of the six corners (Fig. 3.7-2). The ionisation energy of the process important for breakdown, namely:

$$SF_6 \rightarrow SF_5^+ + F^-$$

is 19.3 eV.

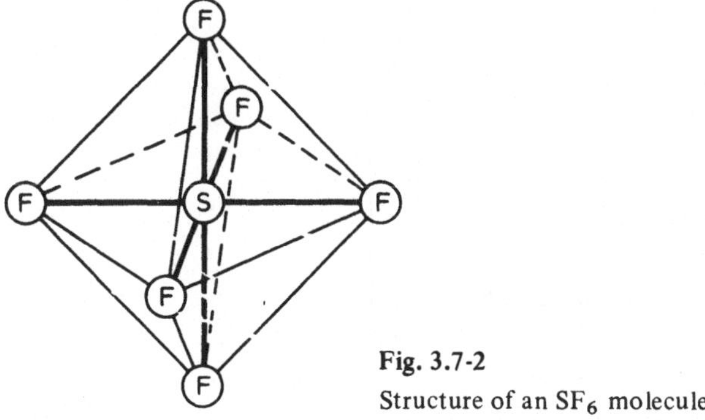

Fig. 3.7-2
Structure of an SF_6 molecule

Sulphurhexafluoride, with density 6.139 g/l at 20 °C and atmospheric pressure, is one of the heaviest gases and is 5 times as heavy as air. It is colourless, odourless, non-toxic and chemically very inactive. Since SF_6 has no dipole moment, ϵ_r is 1 and independent of frequency.

The electrical strength of SF_6 in a homogeneous electric field is 2 to 3 times that of air. The breakdown mechanism of SF_6 is not yet fully explained. As a consequence of the tendency to form negative ions, an appreciable variation of the individual processes would be expected compared with air. Results of measurements show, however, that the discharge growth in SF_6 can also be described reasonably well using the concepts of classical gas breakdown theory. This is shown by the pressure dependence of the breakdown voltage. The transition from the Townsend mechanism to the less advantageous streamer mechanism is expected for a very much lower pressure in SF_6 than in air. This is also especially true for the reduction of U_d in a strongly inhomogeneous field, shown in Fig. 3.7-1b [*Hartig* 1966].

During arc discharges in SF_6 reactive and toxic by-products are formed, which have to be absorbed by suitable agents (e.g. Al_2O_3).

3.7.2 Experiment

The data of the most important circuit elements are:

T	Testing transformer 220 V/200 kV, 10 kVA
CM	Measuring capacitor 200 kV, 100 pF
G	Rotary vacuum pump, model D 6 [1])

a) Experimental Setup

The experiments are performed with the setup shown in Fig. 3.7-3. The alternating test voltage obtained from the test transformer T should be measured using the peak voltmeter SM (see section 3.1) via different measuring capacitors CM. The vacuum necessary for the experiment is generated by the rotary pump G and measured by a membrane vacuum meter M. The regulating valve D is used for exact regulation of the desired pressure. For

1) Manufacturer Messrs. Leybold, Köln

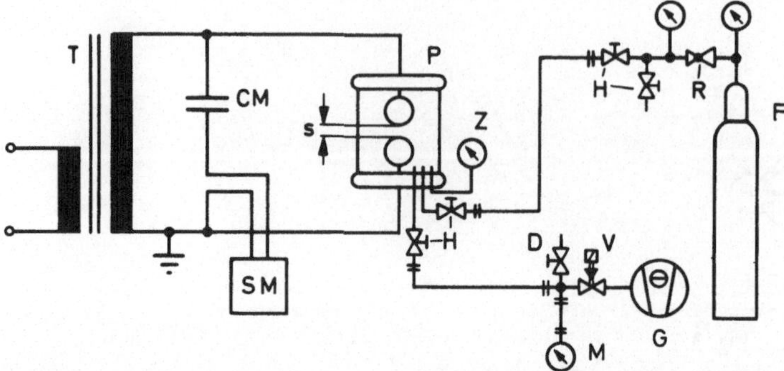

Fig. 3.7-3 Test setup for measurement of breakdown voltage at pressures of 1 mbar to 6 bar

measurements in the high pressure range (**Attention**: limitation owing to the mechanical strength of the pressure vessel!) a gas cylinder F with a reducer valve R should be connected (**Attention**: the gas cylinder should be securely fixed to prevent toppling!). The high pressure is measured using the indicating manometer Z mounted on the pressure vessel. Before beginning the high-pressure experiments, one should make sure that the membrane vacuum meter M is disconnected, to avoid damage. The stop cocks H allow connection of the required pipe-lines; the magnetic valve V closes automatically when the pump is switched off, so that unintentional aeration of the container is prevented.

The testing arrangement P is set up in a pressure vessel as shown in Fig. 3.7-4. The insulating tube is of perspex and thus permits visual observation of the discharge phenomena. The electrodes can be exchanged by means of the removable insets; as an example, the figure shows an arrangement of two spheres of diameter D = 50 mm and spacing s = 20 mm, the most commonly used for the experiments. The pressure vessel is suitable for the proposed pressure range of about 1 mbar to 6 bar and withstands a test pressure of about 10 bar. The ring-shaped grading electrodes shown are necessary for increasing the external flashover voltage. In this way measurements up to 200 kV a.c. could be carried out with this testing vessel.

b) Validity of the Paschen Law for an Electrode Configuration in Air

The electrode system to be investigated is a sphere gap with D = 50 mm. The a.c. breakdown voltage U_d in air should be measured for spacings s = 10 mm and 20 mm. The relation shown in Fig. 3.7-5 was obtained for the described experiment. From this it follows that the conditions of the Paschen law are well satisfied. Furthermore, diverse types of gas discharge occur after breakdown in the investigated pressure range. Fig. 3.7-6 shows a glow discharge at about p = 10 mbar and an arc discharge at normal pressure.

c) Breakdown Voltage of an Electrode Configuration in SF_6

With the aid of a second testing vessel as in Fig. 3.7-4, comparative measurements of the breakdown voltage U_d of the sphere gap should be carried out in SF_6 at spacing s = 20 mm and for a pressure range of 1 to 6 bar. The gas pressure is produced by an SF_6 compressed gas cylinder.

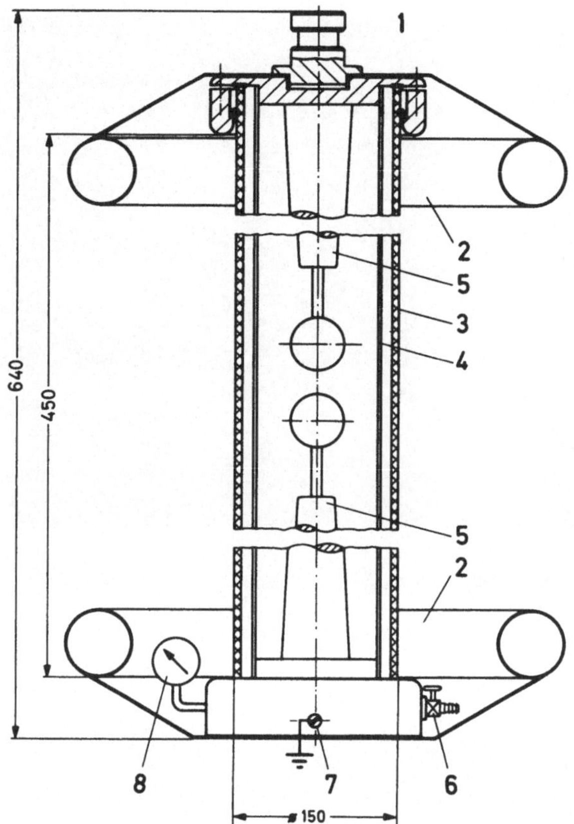

Fig. 3.7-4

Testing vessel for measurement of the breakdown voltage at pressures of 1 mbar to 6 bar

1 Vessel lid (high-voltage terminal)
2 Grading electrodes
3 Perspex cylinder
4 Hardboard cylinder
5 Electrode support
6 Stopcocks
7 Earth terminal
8 Pressure gauge

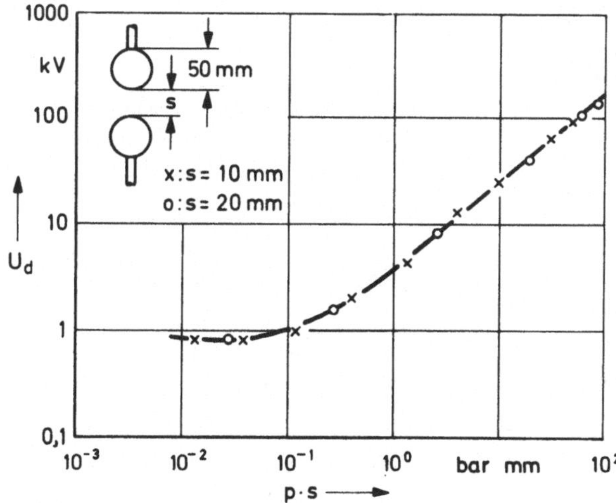

Fig. 3.7-5

Measured values of breakdown voltage between spheres in air

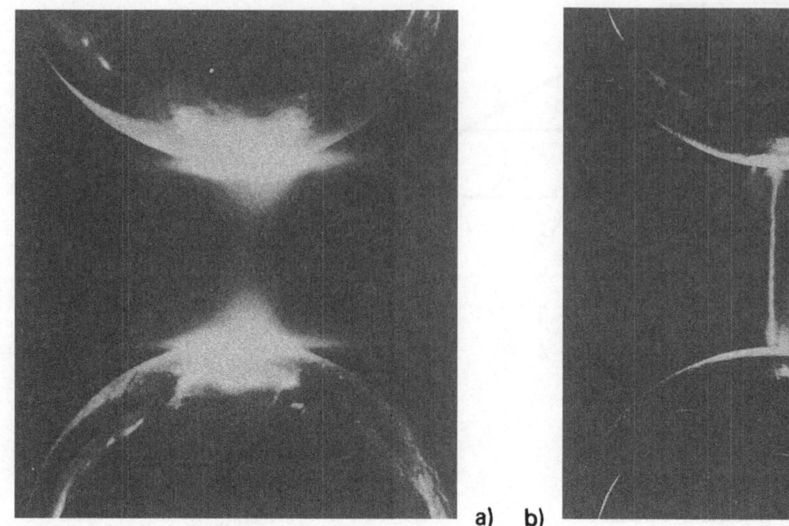

Fig. 3.7-6 Types of gas discharge in air at alternating voltages, spacing s = 20 mm
a) Glow discharge, pressure 10 mbar, exposure 5 s
b) Arc discharge, pressure 1 bar, exposure 40 ms

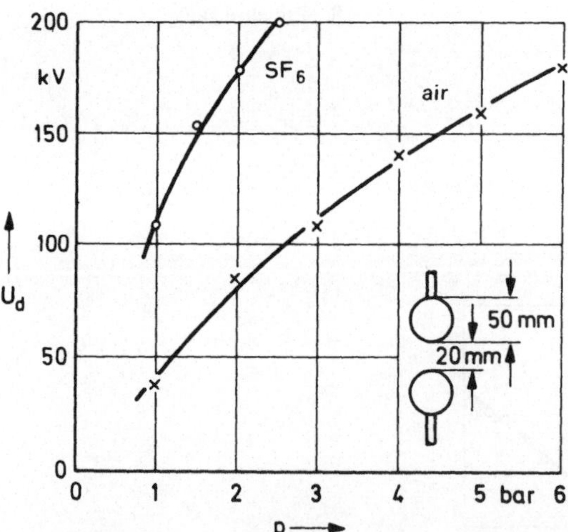

Fig. 3.7-7
Pressure dependence of the breakdown voltage of a sphere gap in air and in sulphurhexafluoride

It is recommended that the measurements in SF_6 and in air be conducted in separate testing vessels, because once a vessel is filled with SF_6 the residual gas would continue to affect the results of later measurements in air, despite long evacuation periods.

For measurements performed with the test system described, the values indicated in the diagram of Fig. 3.7-7 were obtained. At the same pressure, the strength of SF_6 is a factor of 2 greater than that of air.

d) Pressure Dependence of the Breakdown Voltage in a Strongly Inhomogeneous Field

To demonstrate the breakdown performance of SF_6 as a function of pressure in a strongly inhomogeneous field, a point-plane electrode configuration is chosen. The diameter of the plate is D = 50 mm and the point is a 10° cone cut out of a 10 mm diameter rod. The gap spacing s should be set to 40 mm and measurements conducted in the pressure range of 1 ... 6 bar. The equipment for producing and measuring low pressures may therefore be disconnected.

For the above experiments the relation shown in Fig. 3.7-8 was obtained for the spacings s = 20, 30 and 40 mm. The falling tendency of the breakdown voltage at increasing pressures within a certain range, lies at appreciably lower values of pressure for heavy gases such as SF_6 than for lighter gases such as air. This effect may be accounted for by a change in the discharge mechanism, namely by the transition from the Townsend mechanism to the streamer mechanism [*Hartig* 1966].

3.7.3 Evaluation

The breakdown voltages as the function $U_d = f(ps)$ for the sphere gap measured under 3.7.2b at spacings s = 10 mm and 20 mm at different pressures should be represented in a diagram on double-logarithmic paper.

The above values of the breakdown voltages U_d of the sphere gap for s = 20 mm in air, together with the values measured in SF_6 under 3.7.2c, should be represented in a diagram as $U_d = f(p)$.

The breakdown voltages of the point-plane system in SF_6 measured under 3.7.2d should be shown as a function of pressure $U_d = f(p)$.

Literature: *Gänger* 1953; *Meek, Craggs* 1953; *Sirotinski* 1955; *Llewellyn-Jones* 1957; *Flegler* 1964; *Raether* 1964; *Kuffel, Abdullah* 1970

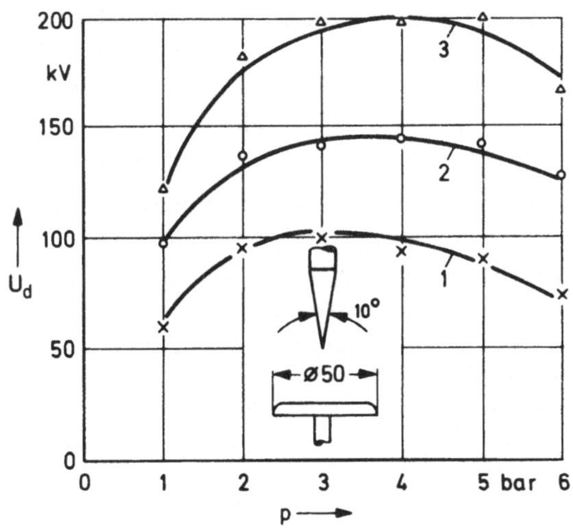

Fig. 3.7-8
Pressure dependence of the breakdown voltage of a point-plane gap in SF_6

1 s = 20 mm
2 s = 30 mm
3 s = 40 mm

3.8 Experiment "Impulse Voltage Measuring Technique"

The time-dependent character of an impulse voltage is often appreciably affected by the properties of the test object connected to the generator. This is particularly true for the case of breakdown phenomena with resultant intentional or unintentional chopping of the impulse voltage. Further, knowledge of the impulse voltage-time curves of practical high-voltage equipment is important for coordination of insulation in systems. Oscillographic measurements of rapidly varying voltages are therefore indispensable to the assessment of test results.

The topics treated in this experiment fall under the following headings:

Multiplier circuit after Marx
Impulse voltage divider
Impulse voltage-time curves

Prerequisite for successful participation is a familiarity with sections:

1.3 Generation and measurement of high impulse voltages
3.3 Experiment "Impulse Voltages"

In this experiment the *Marx* circuit for the generation of impulse voltages shown in Fig. 1.3-4 is used. The data of the single-stage equivalent circuit, enabling calculation of the voltage form, can be determined according to section 1.3.2.

3.8.1 Fundamentals

a) Elements of an Impulse Voltage Measuring System

The block-diagram of a complete impulse voltage circuit is shown in Fig. 1.3-11. The high impulse voltage $u_1(t)$ to be measured must first be greatly reduced by a voltage divider. From a tap on this divider a measuring voltage proportional to the high-voltage signal is fed through a measuring cable, either to a cathode-ray oscilloscope or to an electronic peak voltage measuring instrument. The load capacitor of the impulse generator itself is often concurrently used as a capacitive voltage divider. For an impulse generator set up as in circuit a, the discharge resistor can also be employed as a resistive voltage divider. However, these arrangements are only suitable for the determination of the peak value of a full or tail-chopped lightning impulse voltage of the form 1.2/50. They are less suited for measurement of impulse voltages chopped on the front. A voltage divider which does not need to serve a double purpose, can be adapted better to the requirements of measurement.

Since the voltage at the input of the oscilloscope can have any desired magnitude, an amplifier is not necessary; the voltage can be applied directly to the vertical deflection plates of the oscilloscope. These oscilloscopes are built with deflection sensitivities of 50 ... 150 V/cm. Potential differences within the measuring system and induced disturbances of the order of a few volts for these oscilloscopes only nominally affect the phenomena to be measured. If oscilloscopes with amplifiers are used, then additional earthing and screening measures invariably become necessary.

On its way from the test object terminals to the oscilloscope, the signal to be measured is distorted, namely in general the more so the higher the frequency components it contains.

For measurement of fast-varying impulse voltages therefore, it is essential that the transmission behaviour of the measuring system be checked.

In high-voltage technology the step function response of the entire measuring system is adopted as a measure of the fidelity. A characteristic parameter of the step response is the response time T. For known electrical properties of the divider, it can either be calculated or determined experimentally at low or high-voltages. In this experiment the method of determining T using a test gap of exactly known impulse voltage-time curve, as described under 1.3.7, is applied.

b) Impulse Voltage-Time Curves

When electrode systems are stressed by impulse voltages of a certain form and higher peak values than necessary for breakdown, then these are referred to as overshooting impulse voltages. In these tests, the higher the peak value $\hat{U}$ is of the unchopped full impulse voltage, the shorter the time t_d becomes to breakdown of the test object. These correlations are described by the impulse voltage-time curve $U_d = f(t_d)$, which is typical for a given system and voltage form; whereby U_d is the highest voltage value prior to breakdown and t_d is the time interval between the start of the impulse (point 0_1 in Fig. 1.3-2) and the start of the voltage collapse. For every impulse voltage-time curve one should specify the set impulse voltage form as well as the polarity on which the given characteristic is based.

According to section 3.3, the breakdown of a gap occurs only when a voltage greater than the static onset voltage U_e persists at the gap for periods longer than the sum of the statistical time-lag t_s and the formative time-lag t_a. Since the front-time of a lightning impulse voltage of a given form is independent of the peak value $\hat{U}$, the voltage rise becomes steeper with increasing peak value. Hence for greater steepness the voltage can increase further beyond U_e during the breakdown time-lag t_v; the increase of U_d for higher overshooting voltages is thus explained.

The statistical time-lag and the formative time-lag are not however independent of the applied voltage. In systems with homogeneous or only slightly inhomogeneous fields (example: sphere gap), t_s and t_a decrease rapidly with increasing overvoltage $\hat{U}/U_e$. In a system with strongly inhomogeneous field (example: rod gap), the formative time-lag determines the total time-lag and decreases, even at high overvoltages, only slowly compared with the homogeneous field case. Consequently the drop of the impulse voltage-time curve for short breakdown times is more pronounced for a system with an inhomogeneous field than for one with a homogeneous field.

The experimental determination of the impulse voltage-time curve of a given electrode configuration requires numerous individual measurements with various types of voltage forms. For this reason many authors have tried to determine the impulse voltage-time curve by calculation, using assumptions based on physical reasoning. Investigations have shown that such assumptions are valid only for a limited number of cases [*Wiesinger* 1966; *Hövelmann* 1966]. Even so, calculation of the impulse voltage-time curve under certain restricting conditions and for a particular range of breakdown times is still meaningful; it offers the possibility of converting an impulse voltage-time curve measured for a similar configuration.

For calculation of the impulse voltage-time curve of electrode configurations with a homogeneous or slightly inhomogeneous field, the assumption has proved useful that for a given gap the "formative area", i.e. the voltage-time area F above the static breakdown voltage U_e, remains constant even for different voltage forms [*Kind* 1958]:

$$F = \int_{t_e}^{t_d} [u(t) - U_e]\,dt = \text{constant}.$$

The lower integration limit is fixed here by $u(t_e) = U_e$. These relations are illustrated in Fig. 3.8-1 for stress by linearly rising impulse voltages with various gradients.
If the formative area of a system is known by measurement with a particular voltage form, the breakdown voltage for any other voltage form can be calculated; this is particularly easy for the case of linearly rising impulse voltages. Here, for rate of rise S, one has:

$$F = \frac{1}{2}\frac{(U_d - U_e)^2}{S}$$

$$U_d = U_e + \sqrt{2FS}$$

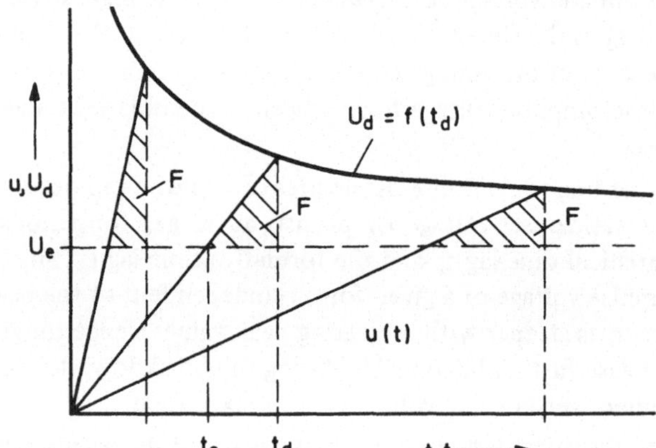

Fig. 3.8-1
Formative area and impulse voltage-time curve

For the test gap mentioned in section 1.3.7 the formative area is $F \approx 2$ kV μs for example.
The statistical variations of the breakdown time-lag have so far been neglected. In practice the impulse voltage-time curve obtained is not a single curve but a band of voltage-time curves, whose upper limit corresponds to a breakdown probability of 100 %, referred to as the statistical time-lag curve. The lower limit is termed formative time-lag curve and corresponds to a breakdown probability of 0 %.
To guarantee effective insulation coordination for overshooting impulse voltages, the impulse voltage-time curve of an overvoltage protective device must lie below that of the equipment requiring protection, for all kinds of voltage gradients. This is generally ensured when surge diverters are used. However, if instead of the diverter a rod gap is used as a protective gap, the safety of the equipment is no longer guaranteed. The band of impulse voltage-time curves of a rod gap rises rapidly with the rate of rise of the voltage, whereas the voltage-time curve of an internal insulation, experimentally determinable only for simple models, can be flat even for very high rates of rise.

3.8.2 Experiment

a) Setup and Investigation of a 2-Stage Impulse Generator

Using the high-voltage construction elements, a two-stage impulse generator should be set up as in circuit a, to generate a positive 1.2/50 impulse voltage. The spatial arrangement recommended for the elements is shown in Fig. 3.8-2.

The elements required have already been mentioned under experiment 3.3. Apart from most of the elements in duplicate, on account of the two-stage construction two additional charging resistors RV = 50 kΩ are required. An oscilloscope with a bandwith of about 50 MHz is suitable. If a special impulse voltage oscilloscope is not available, it is advisable to make use of a measuring cabin (see 2.2).

The impulse generator is triggered by a pulse on F_1. The discharge resistance 2 RE = 19 kΩ, in this circuit parallel to the test object, is used as the high-voltage arm of a resistive voltage divider (divider I). The voltage at the low-voltage resistance arm of this divider is fed to the oscilloscope KO by a coaxial, surge impedance Z = 75 Ω terminated measuring cable.

The faultless operation of the impulse generator, including the oscilloscope KO triggering, should be checked over a large range of trigger gap spacings. Then two full impulse voltage oscillograms with 75 kV charging voltage per stage should be recorded for different time-bases. In addition the peak value of this impulse voltage should be measured with a sphere gap of D = 100 mm.

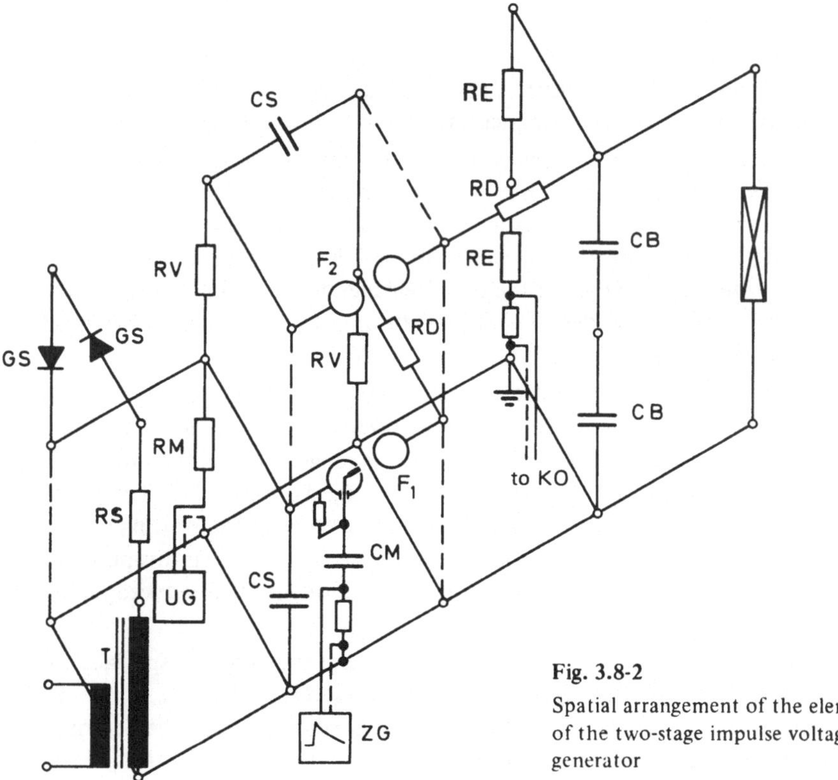

Fig. 3.8-2
Spatial arrangement of the elements of the two-stage impulse voltage generator

For this experiment the oscillogram shown in Fig. 3.8-3 was recorded, from which the time parameters can be taken:

$T_s = 1.23\ \mu s$ and $T_r = 45.6\ \mu s$.

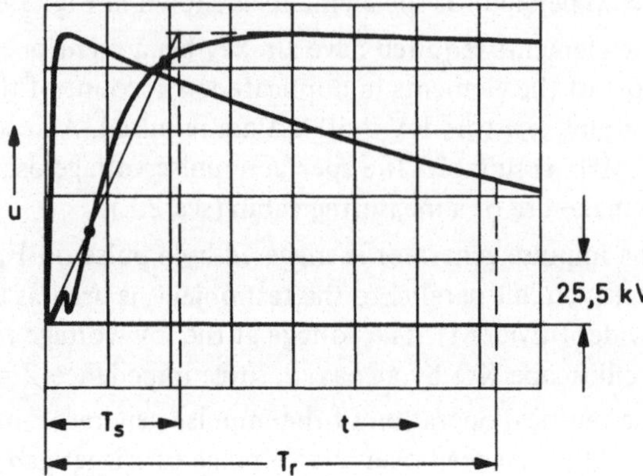

Fig. 3.8-3
Oscillogram of impulse voltage
Time-base: 0.57 and 5.7 μs/division
Charging voltage per stage: 75 kV
$T_s = 1.23\ \mu s$, $T_r = 45.6\ \mu s$

b) Comparison of the Fidelity of Two Impulse Voltage Dividers

A sphere gap with D = 100 mm and s = 30 mm should be used as the test gap for comparing the fidelity of the two voltage dividers, as described in section 1.3.7. In addition to the resistive voltage divider I, a damped capacitive voltage divider according to Fig. 2.4-2 (divider II), the construction of which is described in section 2.4, should also be investigated. When divider II is used, the connections are as in Fig. 1.3-10a.

For each divider the time dependence of the voltage at the gap should be recorded in a common oscillogram, whilst the gap is stressed by three strongly overshooting impulse voltages. In doing so divider II should be connected in parallel to the test gap with a short lead of about 1 m length. Triggering of the oscilloscope KO is done with the aid of an antenna or directly by the trigger pulse for the impulse generator. Fig. 3.8-4 shows the "true" impulse voltage-time curve under standard conditions of the test gap used here. It was calculated by the method outlined in 1.3.7 for F = 1.06 kVμs.

Using the oscillograms the measured breakdown voltage $(U_d)_{gem}$ of the test gap should be determined as a function of the rate of rise S of the voltage. This rise should be approximated as closely as possible by a straight line of slope S in the range between the measured breakdown voltage of the gap and its static breakdown voltage $U_e = 85.5$ kV. The point of intersection of this straight line with the base line shall be taken as the starting point of the idealised measured wedge-shaped impulse voltage. The time between this point and the voltage collapse is the measured breakdown time $(t_d)_{gem}$ of the idealised wedge-shaped impulse voltage.

The pair of values $(U_d)_{gem}$ and $(t_d)_{gem}$ contain amplitude and time errors, especially in the region of large rise. The rate of rise S shall however be assumed to have been measured accurately. The response times of the two voltage measuring systems should be determined by comparison of the measured values with the voltage-time curve of Fig. 3.8-4.

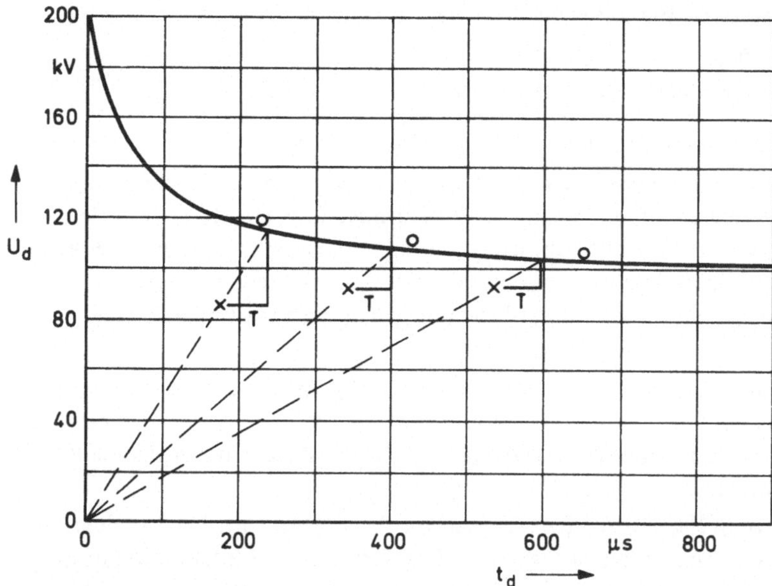

Fig. 3.8-4 Impulse voltage-time curve of the test gap (100 mm diameter, s = 30 mm) for positive wedge-shaped impulse voltages at standard conditions

x – measured points using divider I obtained at d = 0.95 o – measured points using divider II

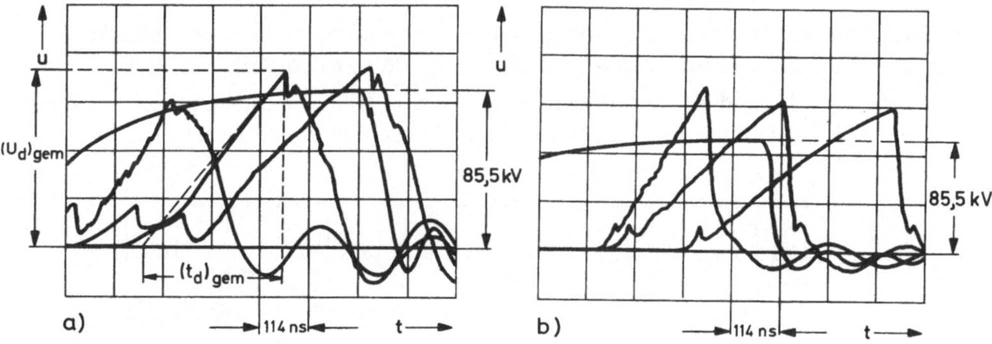

Fig. 3.8-5 Oscillograms of the voltage curves for stressing a sphere gap of s = 30 mm in air, calibration with U_{d-50} = 85.5 kV

a) Resistive divider I b) Damped capacitive divider II

For this experiment the oscillograms shown in Fig. 3.8-5 were obtained for the two dividers. The static breakdown voltage of the test gap was determined in each case with a tail-chopped impulse voltage. The value pairs taken from the oscillograms of breakdown voltages and breakdown times for measurements conducted at d = 0.95, referred to standard conditions, are plotted in Fig. 3.8-4. The corresponding "true" points of the impulse voltage-time curve are the points of intersection obtained when a straight line of slope S passes through these measured points. The response time T and the voltage error ST can be read out directly.

For the example shown here, the following mean values were obtained:

Divider I: T = 60 ns
Divider II: T ≈ 0 (evaluation uncertain).

c) Plotting Impulse Voltage-Time Curves

As in section 3.8.2b the test objects sphere gap with s = 45 mm and support insulator with a protective gap of spacing s = 86 mm should be investigated according to choice. Here too the time dependence of three different overshooting impulse voltages should be recorded for each test object. The sphere gap was chosen for these measurement because its impulse voltagetime curve is similar to that of the internal insulation of high-voltage equipment.

The breakdown voltage of each of the two investigated configurations should be determined from the oscillograms as a function of the breakdown time t_d. This latter shall be computed from the impulse start of the 1.2/50 standard impulse voltage. Divider II should be used for the measurements.

The oscillograms shown in Fig. 3.8-6 were recorded for this experiment. Their evaluation yields the impulse voltage-time curve of Fig. 3.8-7. From the point of intersection of the impulse voltage-time curves of the sphere gap and the protective gap, the value 0.12 kV/ns can be read out as a rough approximation for that rate of rise up to which protection is still provided.

3.8.3 Evaluation

The data of the switching elements for the single stage equivalent circuit of the arrangement as in Fig. 3.8-2 and the utilization factor η should be calculated.

T_s, T_r should be determined according to 3.8.2a, and $\hat{U}$ from the oscillograms; η should be determined and $\hat{U}$ compared with the result of the sphere gap measurement.

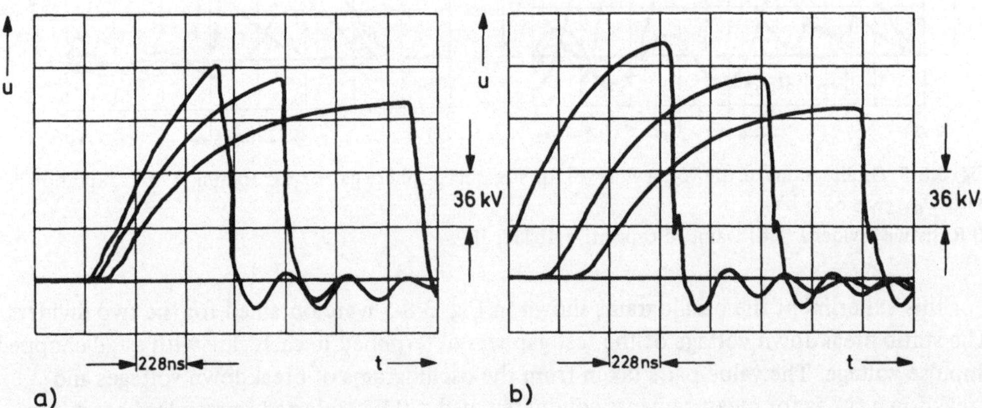

Fig. 3.8-6 Oscillograms of voltage forms for stress of different test objects by overshooting impulse voltages

a) Sphere gap (s = 45 mm) b) Protective gap (s = 86 mm)

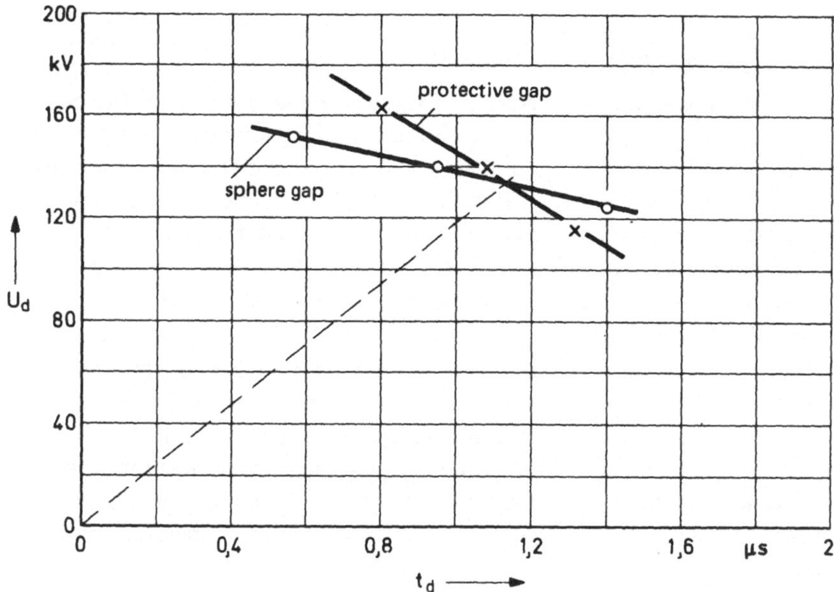

Fig. 3.8-7 Impulse voltage-time curves of a sphere gap (D = 100 mm, s = 45 mm) and a protective gap (s = 86 mm)

The response times of the different dividers as in 3.8.2b should be compared. The impulse voltage-time curves should be plotted as in 3.8.2c.

Literature: *Strigel* 1955; *Schwab* 1972

3.9 Experiment "Transformer Test"

The testing of technical products on the basis of certain specifications serves as a confirmation of agreed properties. Power transformers are important and costly elements in high-voltage networks; their reliable valuation by means of high-voltage tests is therefore of particular significance to the operational security of electrical supply systems.

The topics covered by this experiment fall under the following headings:

Specifications for high-voltage tests
Insulation coordination
Breakdown test of insulating oil
Transformer test with alternating voltage
Transformer test with lightning impulse voltage

It is assumed the reader has some basic knowledge of the construction of 3-phase power transformers and familiarity with the sections:

1.3 Generation and measurement of high impulse voltages
3.3 Experiment "Impulse Voltages".

3.9.1 Fundamentals

a) VDE-Specifications and IEC-Recommendations [1)]

Obligatory guidelines and regulations are necessary for the assessment of the quality and for the trade of electrotechnical products. At the national level in Germany, these are provided and published by the technical committees of the "Verband Deutscher Elektrotechniker" (VDE). In order that they may not inhibit future development, the VDE specifications must be regularly revised and amended, to meet the corresponding standard of technology. In addition to important safety regulations, they also contain instructions for conducting tests. In this way recognized rules of electrotechnology resulted, which, in the event of damages, are also of legal consequence.

The need to publish similar specifications at the international level followed from the increasing expansion of trade beyond national borders. Due to the many differences prevalent amongst the nations on account of historical development, climatic variations and unit systems for example, international agreements can only take the form of skeleton recommendations. They are worked out by the technical committees of the IEC. The harmonization of national specifications is of great significance to the economic cooperation between different countries.

The VDE 0532 "Rules for transformers", besides definitions of terms, contains specifications for the construction and testing of transformers. The high-voltage tests mentioned in these and other equipment regulations are in accordance with the rules concerning the magnitudes of test voltages (VDE 0111) and the generation and measurement of test voltages (VDE 0432).

b) Insulation Coordination

In the field of high-voltage technology the "Guidelines for the design and testing of the insulation of electrical equipment for alternating voltages of 1 kV and above" (VDE 0111) assume a special position, since the test voltages are specified there in a uniform way. Voltage series, corresponding to the standardized nominal voltages, are used to characterize the insulation of equipment.

In the case of external overvoltages the definition of an insulation exempt from every risk is usually impossible, for economic reasons. The test voltage for lightning impulse voltages is therefore chosen so that no breakdowns can occur during operation either within the equipment or across open contacts. For insulation coordination it is essential that the strength of the internal insulation (upper impulse level) lies above the breakdown or flashover voltage of air gaps (lower impulse level). Further, the magnitude of overvoltages occurring must be limited by the use of overvoltage protective equipment (protective level). For lightning impulse voltages the voltage form is defined by 1.2/50.

A test using switching impulse voltages to verify the insulation strength against internal overvoltages has special significance for large spacings in air and strongly inhomogeneous field. Air spacings of insulation systems for operating voltages of over 220 kV should

1) See also the relevant IEE and IEEE specifications where appropriate

therefore be subjected to a corresponding type test. Switching impulse tests could also be effective as routine tests in lieu of a test with excessively high alternating voltages.
A few test voltages and protective levels for transformers in 3-phase systems are given in Table 3.9-1 as examples.

Table 3.9-1 Test voltages and protective levels for transformers in 3-phase systems with neutral points not solidly earthed (Insulation group D)

Voltage series kV	AC test voltage kV	Impulse level lower kV	Impulse level upper kV	Protective level kV
10	28	75	85	40
20	50	125	145	80
30	70	170	195	120
60	140	325	375	235
110	230	550	630	415
220	460	1050	1200	825

c) Testing of Insulating Oils

Power transformers for high voltages contain large quantities of insulating oil for insulation and cooling. Good dielectric properties of the insulating oil are therefore an important prerequisite for perfect insulation of these transformers. Since the breakdown strength of an insulating oil depends appreciably upon its composition, preparation and ageing conditions, its determination is an important part of the high-voltage testing of transformers.

In VDE 0370 "Specifications for oils in transformers, instrument transformers and switchgear", a minimum quality is prescribed for new and used oils under exactly specified testing procedures. The complete testing programme covers the following properties, among others: purity, density, viscosity, breakdown voltage, dielectric dissipation factor and specific volume resistivity.

The breakdown voltage should be measured using a standard testing vessel and alternating voltages of supply frequency. The spherical caps with spacing s = 2.5 mm shown in Fig. 3.9-1 should be chosen as electrodes. The test voltage should be increased from zero at a rate of about 2 kV/s to breakdown. Six breakdown experiments should be conducted for each oil sample. The mean value of the breakdown voltage determined from the 2nd to the 6th measurement may not be less than certain minimum values. These values are 60 kV for new oils in transformers and instrument transformers and up to 30 kV for switchgear; lower values are permissible for equipment in service.

d) Testing of 3-Phase Transformers with Alternating Voltages

In high-voltage equipment with windings, one should distinguish between winding insulation tests and interturn insulation tests. Both tests are conducted as routine tests.

For the winding test at the test voltage U_p the insulation between all the high-voltage windings, and the low-voltage windings connected to the core, is tested as shown in

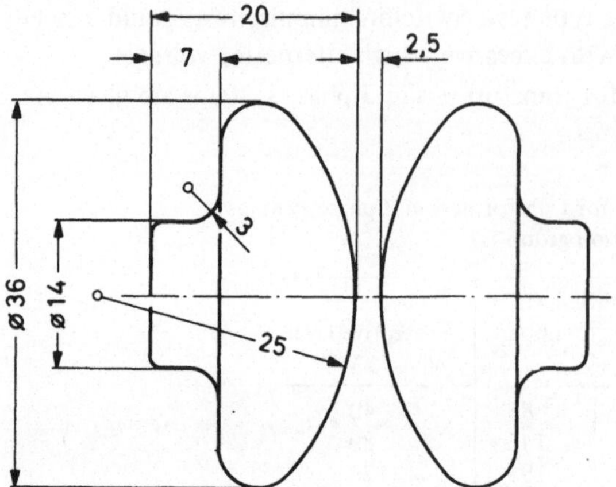

Fig. 3.9-1 Electrodes for the measurement of the breakdown voltage of insulating oils a according to VDE 0370

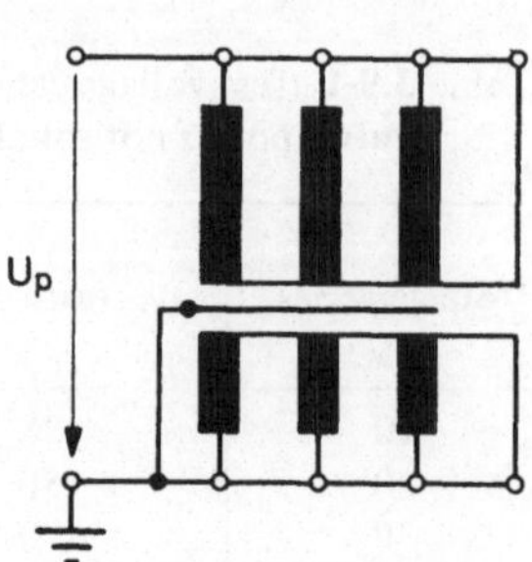

Fig. 3.9-2 Circuit for testing winding insulation

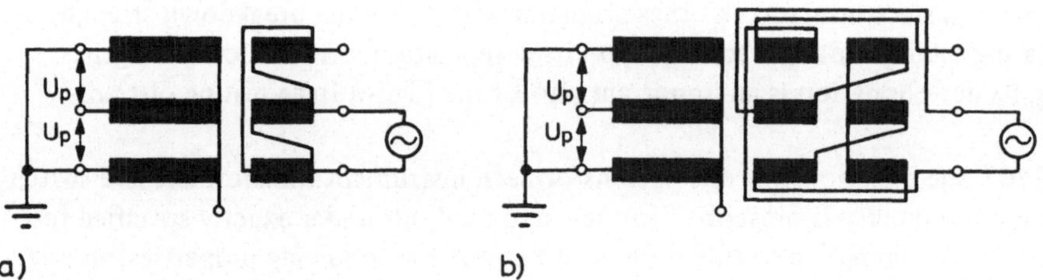

Fig. 3.9-3 Circuits for interturn insulation tests
a) Vector group Yd 5 b) Vector group Yz 5

Fig. 3.9-2. Should the high-voltage windings be single-pole insulated, the winding test on manufactured equipment can be carried out only at a voltage corresponding to the insulation of the earth-side terminal.

For the interturn test (induced voltage test) the mutual insulation of the individual turns is tested. In doing so the testing frequency may be increased, in case the current drawn is excessively large due to saturation of the iron core.

Two circuits are shown in Fig. 3.9-3 for the interturn test of 3-phase transformers with different vector groups. The test should be performed by cyclic interchange of the phases. Excitation is thereby effected by connecting two terminals of the high-voltage or low-voltage winding to an adjustable alternating voltage.

e) Testing of Transformers with Lightning Impulse Voltage

For impulse voltage tests on transformers it is primarily the interturn test which is important, since an uneven voltage distribution along the winding may be anticipated (see also section 3.11). The particular difficulty of this test lies in the reliable identification even of small and only transient partial defects. On no account may a defect develop during the test which remains unidentified and could cause failure in service later on. As a rule, impulse voltage tests on transformers are conducted as type tests.

Fig. 3.9-4 shows a measuring circuit suggested by *R. Elsner* in 1949. The current i_c, which for fast processes is mainly capacitively transferred to the low-voltage winding US, is measured oscillographically by the voltage drop it causes across the measuring resistor R_i. Partial breakdowns in the high-voltage winding OS modify the oscillations induced by the impulse and are further observed by the superposition of a higher frequency oscillation. Defects in the high-voltage winding are identified by comparison of the charging current oscillograms obtained on testing with an impulse voltage low enough not to cause a defect (calibration impulse), and on stressing with the full test voltage (test impulse).

A measuring circuit proposed by *J. H. Hagenguth* in 1944 is shown in Fig. 3.9-5. Here the magnetization current i_0 flowing from the stressed winding to earth is measured. Fault identification again occurs by comparison of the oscillograms obtained during calibration and test impulses.

These tests are generally conducted with full impulse voltages. In special cases a test with chopped impulse voltages can be additionally agreed upon with the customer. Because of the rapid voltage collapse this test represents an especially high stressing of the insulation.

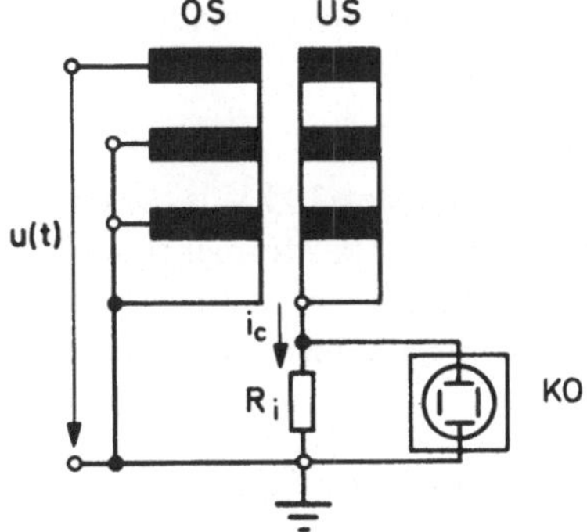

Fig. 3.9-4
Circuit for the lightning impulse voltage test according to Elsner
OS High-voltage winding
US Low-voltage winding

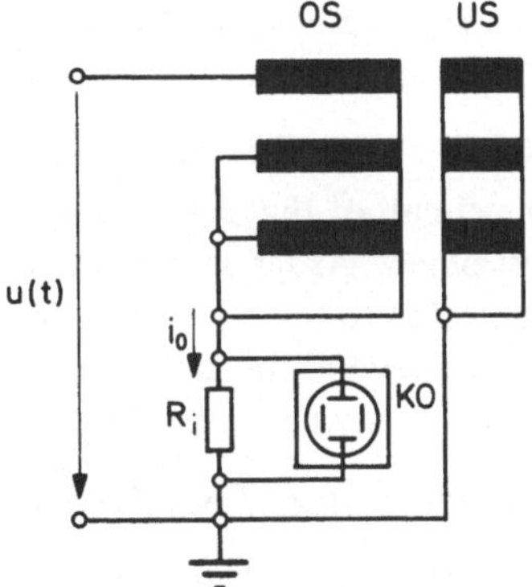

Fig. 3.9-5
Circuit for the lightning impulse voltage test according to Hagenguth
OS High-voltage winding
US Low-voltage winding

3.9.2 Experiment

a) Breakdown Test of an Insulating Oil According to VDE 0370

A circuit as shown in Fig. 3.9-6 should be set up. The following circuit elements will be used:

T Test transformer, rated transformation 220 V/100 kV
CM Measuring capacitor, 100 pF
SM Peak voltmeter (see 3.1)

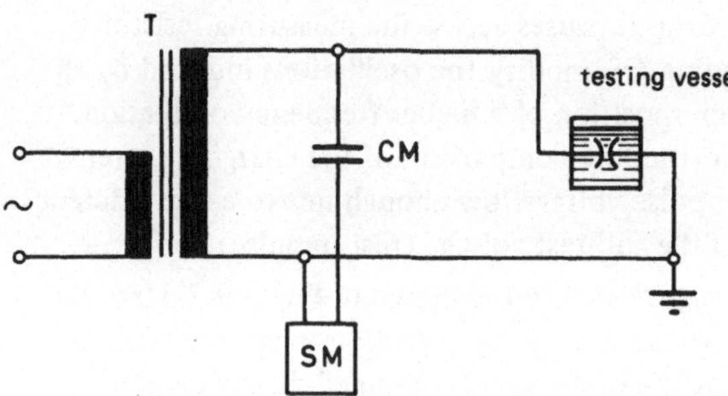

Fig. 3.9-6
Circuit for the breakdown test of insulating oils

An oil sample is taken from the transformer to be tested. The oil to be investigated should be poured slowly into the testing vessel, avoiding bubble formation (by allowing it to run along a glass rod), and then left to stand for about 10 minutes before the voltage is applied. The voltage should be switched off at the instant of breakdown. An interval of about 2 minutes should be held after each breakdown and the breakdown path between the electrodes flushed with new oil by carefully passing a stirring-rod through the gap.

b) AC Test of an Oil-Filled Transformer According to VDE 0532

In the circuit of Fig. 3.9-6, an oil-filled transformer of the voltage series 20 kV is connected as test object. As far as is practicable, the high-voltage winding of the transformer should be subjected to the a.c. test voltages. However, as specified for repeat tests on transformers in service beyond the guarantee period, only 75 % of the test voltage values according to VDE 0111 may be applied here.

c) Impulse Voltage Test of an Oil-Filled Transformer According to VDE 0532

A single-stage impulse generator as in Fig. 3.3-2, but to generate negative lightning impulses of the form 1.2/50, should be set up as in circuit a, Fig. 1.3-3. The 3-phase transformer of 3.9.2b should be connected as the test object, with a rod gap of adjustable spacing s in parallel. Fault identification should be realized with the help of either of the circuits shown in Figs. 3.9-4 and 3.9-5 (guiding value for $R_i = 75\ \Omega$), using a double-beam oscilloscope (bandwidth $\geq$ 1 MHz). The satisfactory working order of the setup, including the oscillographic measurement, should be checked without the test object for d.c. charging voltages $U_0 = 70 \dots 130$ kV. The measurement of the peak value can be effected here via U_0 whereby a constant value is assumed for the utilisation factor η. The test object should then be connected. For verification of the upper and lower impulse levels, the following tests should be performed (though here as repeat tests at only 75 % of the values as new):

2 calibration impulses with full impulse voltages at 75 % of the lower level values
2 test impulses with chopped impulse voltages at the upper level values
2 control impulses with full impulse voltages at 100 % of the lower level value

In doing so, voltage and current oscillograms should be recorded. Before connecting the test object, the spacing of the rod gap should be adjusted so that the impulse voltage at the upper level is chopped after about 2 ... 4 μs. The oscillograms reproduced in Fig. 3.9-7 were obtained for this kind of test. The conformity of the recordings in a) and c) shows that the transformer has passed the test.

3.9.3 Evaluation

The breakdown voltage of the oil investigated in section 3.9.2a should be determined. Can this oil be used in new transformers?

Has the oil-filled transformer passed the a.c. test according to section 3.9.2b?

What is the voltage taken by the neutral point of the high-voltage winding of a transformer during an interturn insulation test (Fig. 3.9-3) at 100 % test voltage? What is the test frequency then approximately required?

By comparing the oscillograms recorded under 3.9.2c one may determine whether the test was satisfactory.

Literature: *Wellauer* 1954; *Strigel* 1955; *Sirotinski* 1965; *Heller, Veverka* 1968; *Greenwood* 1971

3.10 Experiment "Internal Overvoltages"

After switching operations as well as due to non-sustained earth faults and short-circuits in electrical systems, overvoltages caused by transient phenomena may occur; as a result the insulation of working equipment is in danger. Besides these transient overvoltages, permanent overvoltages too can occur in networks whose neutrals are not earthed. Noteworthy here are the power frequency oscillations in inductances with non-linear characteristics, which are termed "ferroresonance".

The topics covered in this experiment can be summarised under the following headings:

Neutral shift
Earthing coefficient
Magnetization characteristic
Jump resonance
Subharmonic oscillations

It is assumed the reader has some basic knowledge of multi-phase systems and the properties of oscillatory circuits.

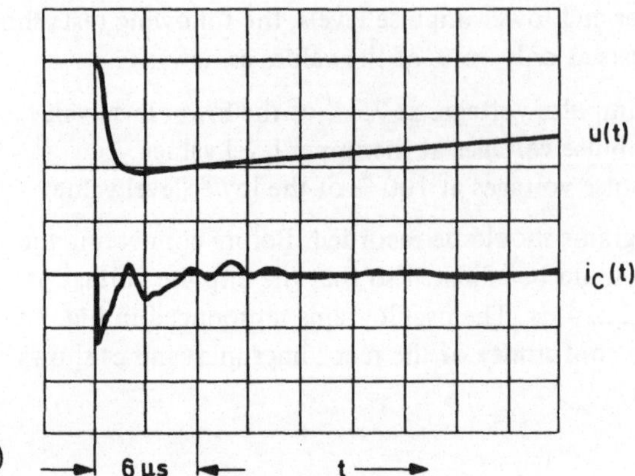

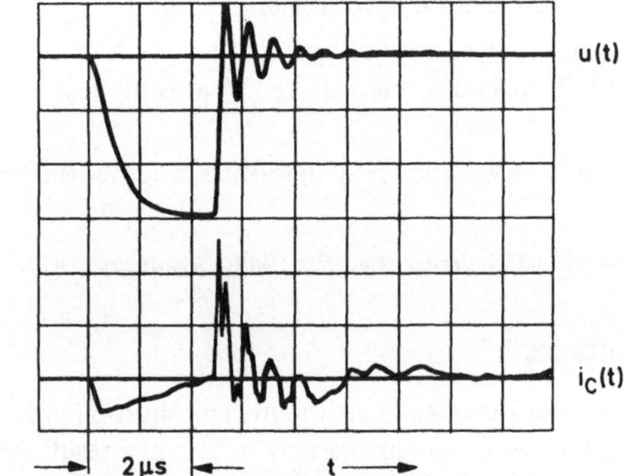

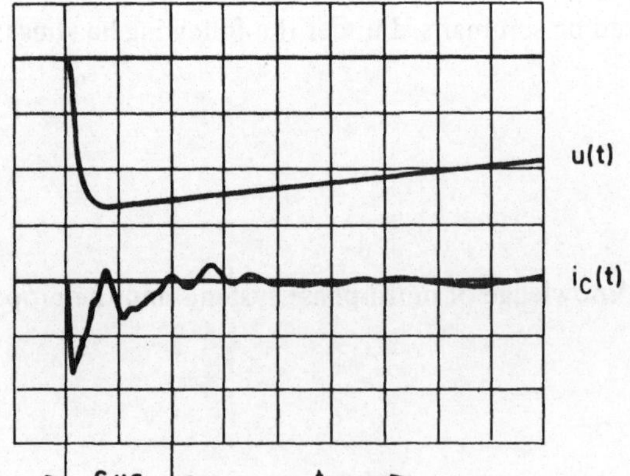

Fig. 3.9-7
Oscillograms of an impulse voltage test on an oil-filled transformer of the voltage series 20 kV according to Elsner

a) Calibration impulse
b) Test with chopped impulse voltage
c) Control with full impulse voltage

3.10.1 Fundamentals

a) Multi-Phase Networks with Non-Earthed Neutral

In 3-phase high-voltage transmission systems with non-earthed neutral point, additional overvoltages can occur as opposed to systems with earthed neutral point. They arise as a result of an earth fault of a conductor or are incited by switching operations as jump resonance and subharmonic oscillations. The reasons for these phenomena may be traced to inductances with non-linear behaviour. These could, for example, as main inductances of voltage transformers or power transformers, form oscillatory circuits with the earth capacitances of the network.

Overvoltage phenomena of this kind occur in a 2-phase network with isolated neutral in a manner analogous to a 3-phase system. Since 2-phase network simulation is easier to realize and the processes during the generation of the overvoltage can be readily surveyed, a 2-phase system is used in the experiment discussed here. Fig. 3.10-1 shows the equivalent circuit valid for basic studies of a 2-phase network, with the transformer T, neutral capacitance C_0, earth capacitances C_{11} and the single-pole insulated voltage transformers W_1 and W_2 with terminal markings U–X and u–x for the primary and secondary windings respectively. The potential of earth is designated by E and that of the midpoint of the transformer by Mp.

For normal working conditions the potential of Mp is determined by the value of the earth and neutral capacitances. Under symmetrical loading conditions and with equal earth capacitances C_{11}, no potential difference exists between Mp and earth E. The potential curves and phasor diagram of a 2-phase system for this working condition are drawn in Fig. 3.10-2. The phase voltages are displaced by 180° with respect to one another. The voltages u_{RO} and u_{SO} can be measured at the terminals u–x of the voltage transformers. When the symmetry of the system is disturbed the potential curves change. Fig. 3.10-3 shows the voltages which would appear on earth fault of phase S. The transformer voltages u_{RO} and u_{SO} are fixed. As a result of the short-circuit the midpoint Mp

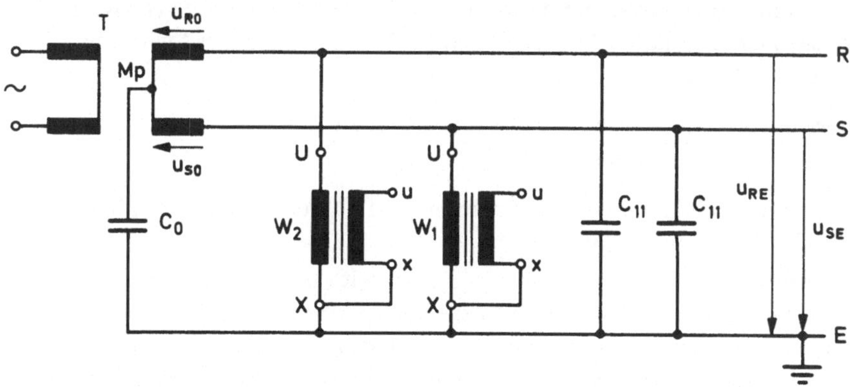

Fig. 3.10-1 Simplified representation of a 2-phase network with non-earthed neutral

T Transformer
C_0 Neutral capacitance
C_{11} Earth capacitances
W_1, W_2 Voltage transformers

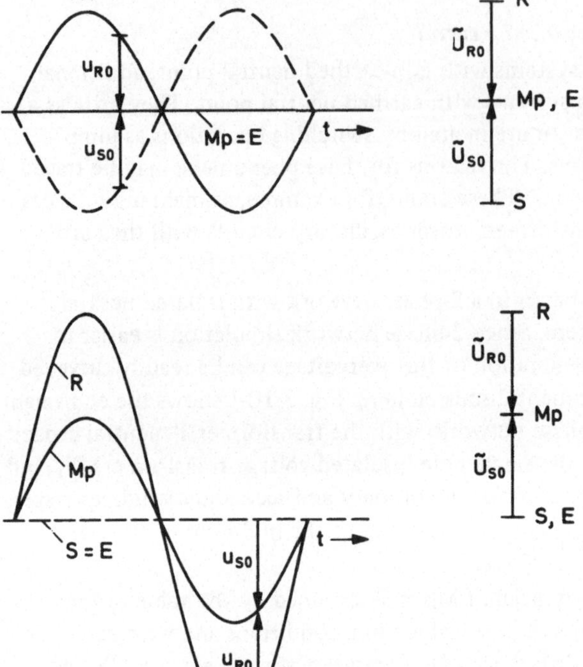

Fig. 3.10-2
Potential curves and phasor diagram of a symmetrical 2-phase network

Fig. 3.10-3
Potential curves and phasor diagram of a 2-phase network with an earth fault in phase S

is displaced with respect to earth E. At the terminals u–x of the voltage transformer of phase S the value measured is zero, and for phase R it is $u_{RO} - u_{SO}$.

In the case of an earth fault the insulation of the intact conductor as well as that of the midpoints of the transformers with respect to earth will be heavily stressed. To give an idea of this voltage increase, the earthing coefficient ϵ has been introduced with the following definition (VDE 0111):

$$\epsilon = \frac{\text{Voltage of intact conductor to earth for earth fault}}{\text{Line-to-line voltage}}.$$

For a multi-phase network with non-earthed neutral $\epsilon = 1$. A network with $\epsilon \leqslant 0.8$ is considered by definition to be solidly earthed.

The network shown with floating neutral in Fig. 3.10-1, with the capacitances C_0 and C_{11} and the current-dependent main inductances of the voltage transformers, represents a non-linear oscillatory circuit. Switching operations or fault conditions can induce the system to oscillate either by way of jump resonance or subharmonic oscillations, depending on the magnitudes of C_0 and C_{11}. An oscillation of the midpoint Mp with respect to the earth potential E also occurs at the same time.

b) Jump Resonance

For closer investigation of the natural oscillations the circuit of the 2-phase network shown in Fig. 3.10-1 is converted to the equivalent circuit as in Fig. 3.10-4. For this purpose the network is assumed to be a linear active four-pole, at the terminals of which the voltage transformers W_1 and W_2 are connected. According to the theory of the equivalent voltage source the capacitance of the 2-phase equivalent circuit is:

$$C^* = C_0 + 2\,C_{11}\,.$$

The analogue treatment of the 3-phase network results in an effective equivalent capacitance between Mp and E:

$$C^* = C_0 + 3\,C_{11}\,.$$

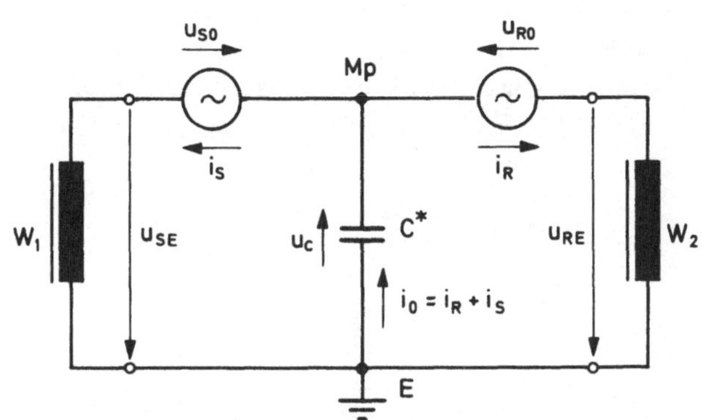

Fig. 3.10-4
Equivalent circuit of the 2-phase network shown in Fig. 3.10-1
C^* Neutral capacitance of the equivalent circuit
W_1, W_2 Voltage transformers

The voltages U_S and U_R remain unchanged in the equivalent circuit, having the same magnitude and opposite phase. For given elements, the currents and voltages for the stationary condition can be approximately calculated [*Philippow* 1963].

With the currents and voltages shown in the equivalent circuit, the following loop equations are valid:

$$u_{RE} + \frac{1}{C^*}\int (i_R + i_S)\,dt = u_{RO}\,,$$

$$u_{SE} + \frac{1}{C^*}\int (i_R + i_S)\,dt = u_{SO}\,.$$

The network with two non-linear elements shown in Fig. 3.10-4 can execute natural oscillations in various ways. The solutions of the system of non-linear differential equations possess great diversity, even when only rough approximations can be made for the characteristics of the two inductances [*Knudsen* 1953; *Philippow* 1963; *Hayashi* 1964]. Hence only three simple limiting cases of the currents i_R and i_S, for which clear-cut solutions can be given, shall be considered here as follows.

$i_R = - i_S$: Phase opposition of currents is to be expected in a fully symmetrical circuit ($u_{RO} = - u_{SO}$). The effective capacitance between Mp and E always remains current free ($i_0 = 0$). Natural oscillations do not occur. This is the symmetrical operating condition.

$i_R = i_S$: For in-phase currents, $i_0 = 2\,i_R$ and the first loop equation takes the form:

$$u_{RE} + \frac{1}{C}\int i_R \, dt = u_{RO}$$

where $C = \frac{1}{2}\,C^*$.

$i_R \gg i_S$: For strong asymmetry of the two loops, $i_0 \approx i_R$ and the first loop equation takes the same form as for $i_0 = i_R$, but with $C = C^*$.

In fact, the condition for complete symmetry is only insufficiently fulfilled solely because of the different characteristics of the voltage transformer inductances; the more the characteristics of the magnetic cores diverge, the greater the current i_0. In this way, as a result of voltage increase or switching operations, the circuit can be induced to carry out neutral oscillations. The loop equation for the two limiting cases mentioned last is satisfied by the series oscillatory circuit shown in Fig. 3.10-5, which is therefore suited for simple investigations of the oscillatory behaviour of networks with non-earthed neutral [*Knudsen* 1953; *Rüdenberg* 1953; *Philippow* 1963; *Hayashi* 1964; *Sirotinski* 1966].

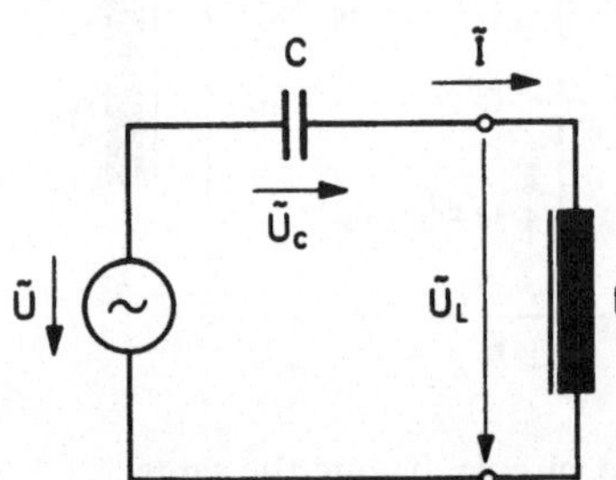

Fig. 3.10-5
Undamped series oscillatory circuit with non-linear inductance

In the following only the fundamental oscillations of the current and the voltage shall be considered. Here the phasor $\tilde{U}$ represents the system voltage and $\tilde{U}_L$ is the voltage across the inductance L. The magnitudes of the phasors are r.m.s. values; this will not be specifically pointed out each time.

The behaviour of the single-phase oscillatory circuit will be explained with the help of the current-voltage characteristic shown in Fig. 3.10-6. For this purpose let us imagine that the circuit of Fig. 3.10-5 is interrupted at the terminals of L. Possible currents $\tilde{I}$ are characterized by the condition that the voltage supplied by the circuit $\tilde{U}_{L1} = \tilde{U} - \tilde{U}_c = \tilde{U} - \tilde{I}/j\omega C$, corresponds to the voltage $\tilde{U}_{L2}$ across the inductance L, which is given by the r.m.s. characteristic.

For the working points A, B and C we have therefore:

$$\tilde{U}_{L1} = \tilde{U}_{L2} = \tilde{U}_L\,.$$

For clarity the diagram also shows the voltages at the working points as phasors.

Investigation of the stability of the working points can be effected by the method of virtual displacement. For an imaginary increase of the current at the point A, $\tilde{U}_L + \tilde{U}_C > \tilde{U}$. Since the driving voltage $\tilde{U}$ is smaller than the sum of the voltages appearing across L and C

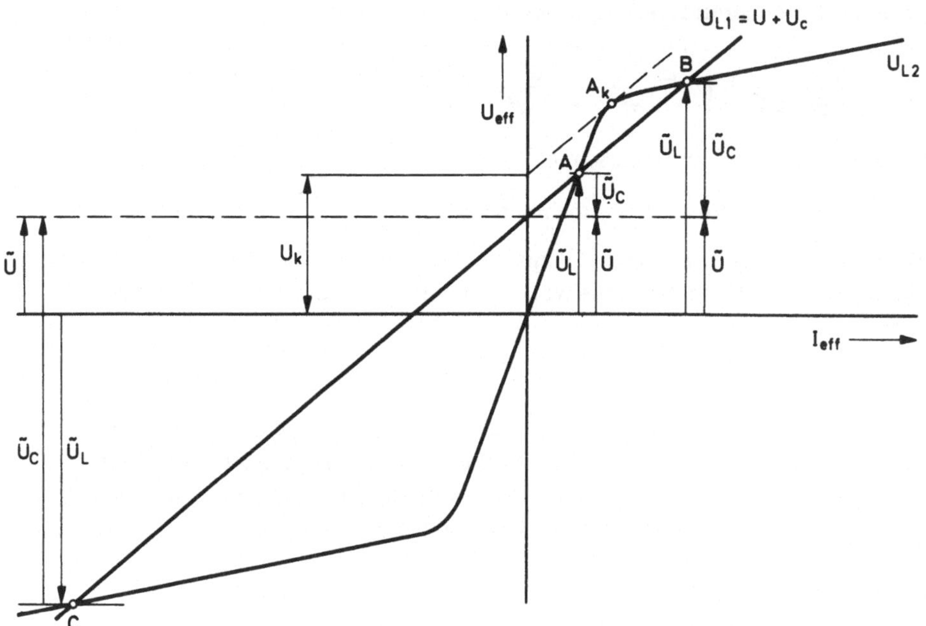

Fig. 3.10-6 Graphical determination of the working points for the oscillatory circuit of Fig. 3.10-5

as a result of the increased current, the current reverts to its original value. The same result is obtained for an imaginary reduction of the current; A is therefore a stable working point. At point B on the other hand, every displacement of the current results in a voltage difference, which strives to increase the deviation further. The condition at B is therefore unstable.

In the same way, it can be shown that C is another stable working point. Owing to the very much higher values of current and voltage compared with those at A, however, operation at point C would mean risk to circuit elements due to overvoltages and thermal overloading.

The jump resonance procedure can be triggered by switching operations, which cause a temporary increase of the supply voltage to $U > U_k$ for example. Jump resonance can also be induced by switching off circuit components and by the associated reduction of the earth capacitance C_{11}; in this event the characteristic $\tilde{U}_{L1}$ in Fig. 3.10-6 would become steeper. The most important means of preventing jump resonance in networks is by introducing damping resistances in the secondary circuit of the voltage transformer, preferably connected to earth fault windings in series. In the single-phase equivalent circuit these resistances R act in parallel with L. This equivalent circuit and the corresponding phasor diagram are shown in Fig. 3.10-7. For the magnitudes we have:

$$U^2 = \left(U_{L1} - \frac{I_L}{\omega C}\right)^2 + \left(\frac{U_{L1}}{R\omega C}\right)^2 = U_{L1}^2 \left(\frac{1+\alpha^2}{\alpha^2}\right) - 2\,U_{L1}\,\frac{I_L}{\omega C} + \left(\frac{I_L}{\omega C}\right)^2$$

with $\alpha = R\omega C$.

The solution of this quadratic equation gives:

$$U_{L1} = \frac{\alpha^2}{1+\alpha^2}\,\frac{I_L}{\omega C} \pm \sqrt{\frac{\alpha^2 U^2}{1+\alpha^2} - \frac{\alpha^2}{(1+\alpha^2)^2}\left(\frac{I_L}{\omega C}\right)^2}\,.$$

The square-root expression represents an ellipse in the current-voltage plane with the semi axes $U\omega C\sqrt{1+\alpha^2}$ and $U\alpha/\sqrt{1+\alpha^2}$; the expression $\frac{\alpha^2}{1+\alpha^2}\,\frac{I_L}{\omega C}$ is a straight line. The sum of the two functions is the sheared ellipse shown in Fig. 3.10-8. The points of intersection A, B and C, as well as the stable working points A and C are again characterized by the relationship:

$$U_{L1} = U_{L2} = U_L\,.$$

For $R \to \infty$ the ellipse degenerates into two straight lines with slope $1/\omega C$ and points of intersection with the ordinate at $\pm U$. With decreasing R the ellipse becomes smaller, corresponding to higher damping, and rises less steeply, so that finally only one stable working point A remains and jump resonance can no longer occur.

c) Subharmonic Oscillations

A further consequence of non-linear inductances in multi-phase systems with floating neutral is the occurrence of subharmonic oscillations. These are stationary oscillations whose frequency is an integral fraction of the supply frequency. In 50 Hz systems, frequencies of 25 and $16\frac{2}{3}$ Hz occur most commonly.

Although these phenomena are very complicated and difficult to calculate, an attempt shall be made here to give a clear explanation of the mechanisms which, to a limited extent, may even be quantitatively interpreted.

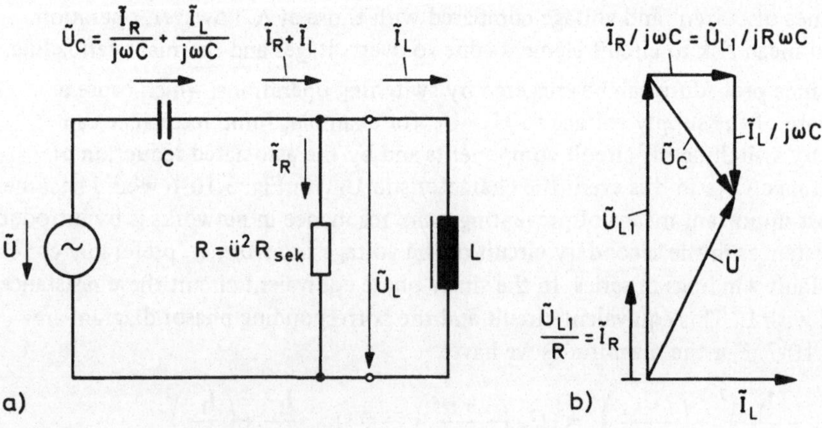

Fig. 3.10-7 Damped series oscillatory circuit with non-linear inductance
a) Equivalent circuit b) Phasor diagram

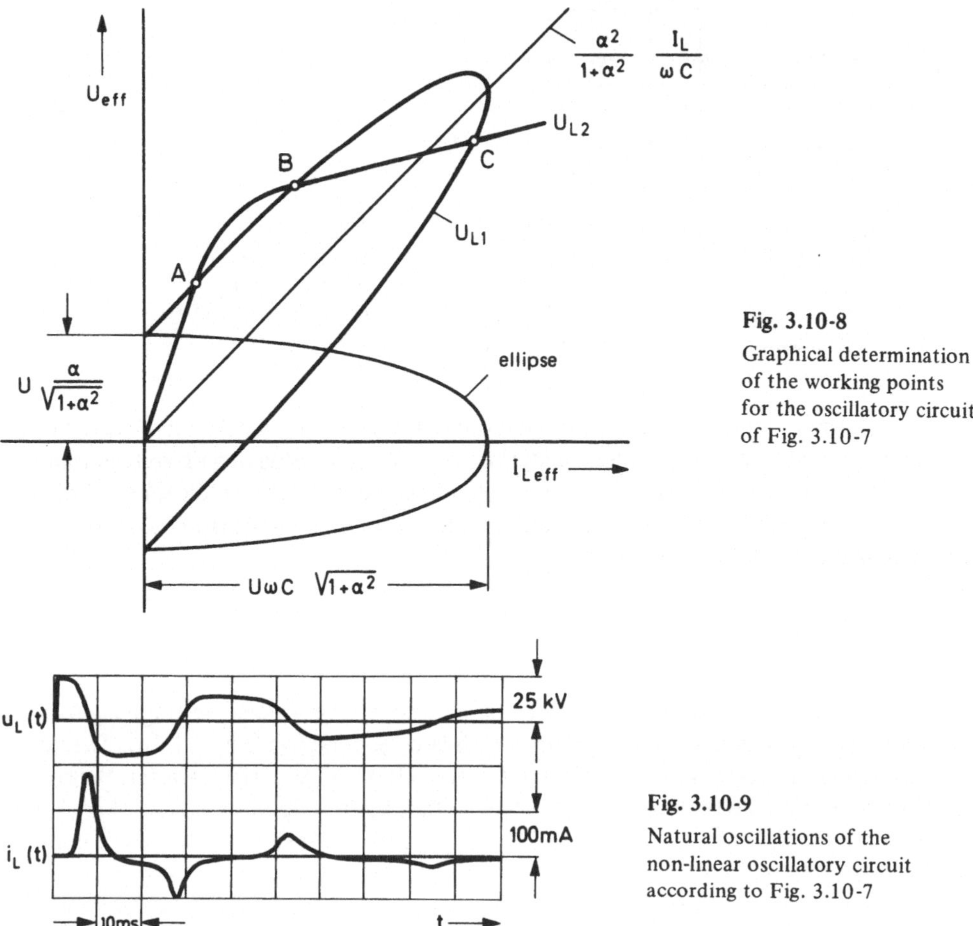

Fig. 3.10-8
Graphical determination of the working points for the oscillatory circuit of Fig. 3.10-7

Fig. 3.10-9
Natural oscillations of the non-linear oscillatory circuit according to Fig. 3.10-7

If the voltage transformer used in section 3.10.2a is connected to a direct voltage source as in Fig. 3.10-7a, the curves of the voltage u_L (t) and current i_L(t), recorded in the oscillogram of Fig. 3.10-9, are obtained.

Since for saturation a constant voltage-time area is always required, the period increases with decaying amplitude.

Such inharmonic oscillations can be induced in the network by switching operations, temporary earth faults or also by jump resonance. Because of damping in the circuit the amplitude of the oscillation fades away and the natural frequency f_e, depending upon the excitation intensity, can traverse a wide range. Synchronisation of the subharmonic oscillation with the 50 Hz supply frequency oscillation is possible, for example when f_e passes through the value 25 Hz and both oscillations simultaneously assume a favourable phase relation with respect to each other. The oscillogram of Fig. 3.10-10 shows subharmonic oscillations of 25 Hz produced in the 2-phase network of Fig. 3.10-4 after interruption of an earth fault in phase R.

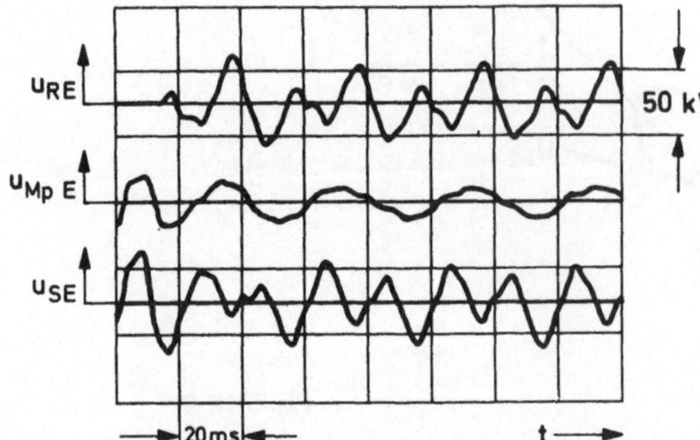

Fig. 3.10-10
Second subharmonic oscillation in a 2-phase network ($f_{1/2}$ = 25 Hz)

In the same way as jump resonance, the subharmonic oscillations cause overvoltages in the system, which can lead to operational disturbances, e.g., by overloading of voltage transformers. By connecting resistances to the series connected earth fault indicating windings of single-pole voltage transformers, the inharmonic oscillations can usually be damped so heavily that they no longer occur.

3.10.2 Experiment

a) Test Setup

The 2-phase network shown in Fig. 3.10-1 can for example be realized with the aid of a test transformer possessing two symmetrical high-voltage windings (Fig. 1.1-1b). The usual adjustable single-phase a.c. supply serves for excitation. As a rule however, these test transformers are not completely insulated, so that their midpoint Mp must therefore be connected to the low-voltage winding at earth potential. But since just the potential shifts of Mp are to be investigated in this experiment, these transformers must be energized via a special insulating transformer. The following equipment was used in the course of this experiment:

T Test transformer 220 V/2 × 50 kV, 5 kVA
Insulating transformer 220/220 V, 50 kV

Two identical single-pole insulated inductive voltage transformers are chosen as high-voltage inductances with non-linear characteristics. The following were used:

$$W_1, W_2 \text{ voltage transformers } \frac{25\,000}{\sqrt{3}} \Big/ \frac{100}{\sqrt{3}} \text{ V.}$$

Before the actual experiment the characteristic $U_{eff} = f(I_{eff})$ of the voltage transformers should be determined, since this will be required for the construction of the diagrams. During these measurements the current must be measured using a true r.m.s. value meter, since its shape deviates strongly from the sinusoidal form; an instrument with a moving-iron mechanism for example is suitable.

To reproduce the capacitance $C = C_0 + 2\,C_{11}$ effective between Mp and E, the construction elements CB and CM, for example, of the high-voltage construction set can be used. The processes to be investigated develop relatively slowly, so that for the oscillographic measurements a bandwith of 8 kHz is sufficient. A 4-beam KO with storage tube was connected to a capacitive voltage divider, with CM = 100 pF as the high-voltage capacitor, and to the secondary terminals of the voltage transformer.

b) Earth Fault Overvoltage in a 2-Phase Network

Using the elements described a circuit of the type shown in Fig. 3.10-1 should be set up. Only CM should be connected between Mp and E as a concentrated capacitance. For symmetrical operation with a phase voltage $U_{RE} = U_{SE} = 10$ kV, earth faults are simulated by temporarily earthing a phase or by short-circuiting the secondary windings of one of the voltage transformers. The voltages

u_{MpE}, u_{RE}, u_{SE} and u_{RS}

should be oscillographed.

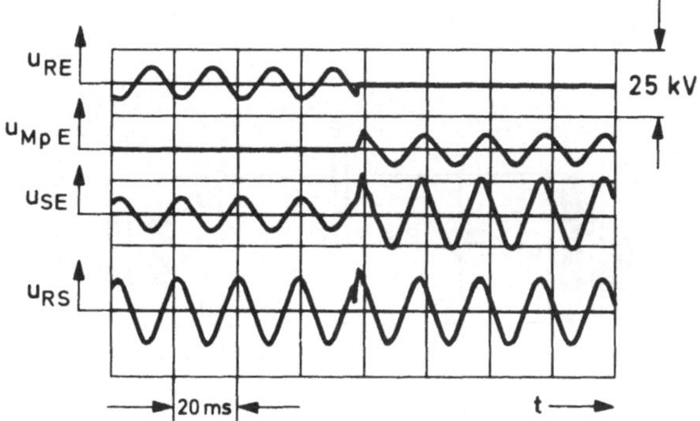

Fig. 3.10-11
Voltage curves for earth fault in phase R in a 2-phase network

During a performance of this experiment the curves shown in Fig. 3.10-11 resulted for earth fault in phase R. As can be seen, appreciable transient and stationary overvoltages appear at Mp and S.

c) Jump Resonance in a 2-Phase Network

In the reconstruction of a 2-phase network an additional concentrated capacitance of 1000 ... 3000 pF is connected between Mp and E. The voltage should be gradually increased under oscilloscopic observation, to the onset of jump resonance.

A voltage lower than that which leads to jump resonance is then applied and jump resonance incited by interruption of the earth fault in a phase.

The oscillograms of Fig. 3.10-12 show the most important curves during the onset of jump resonance. The resonance was caused here by change in the midpoint capacitance.

d) Jump Resonance in a Series Oscillatory Circuit

For these investigations the 2-phase network is reproduced by a series oscillatory circuit as in Fig. 3.10-5. The circuit used is shown in Fig. 3.10-13. As test transformer T the instrument denoted by T above should be chosen, whereby only one high-voltage winding should be connected. An insulating transformer is no longer necessary. A value of 1000 ... 2000 pF is suitable for the capacitance C.

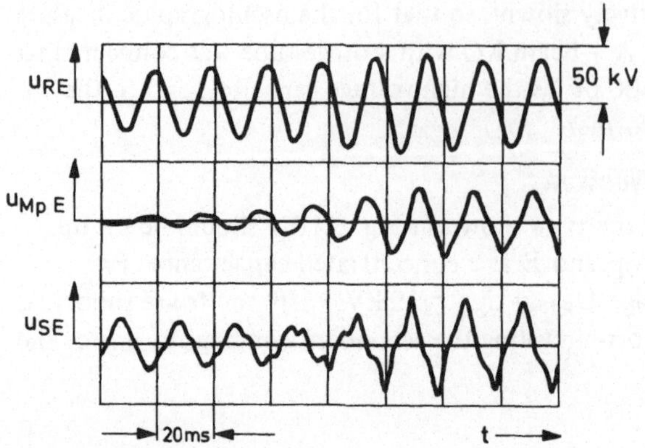

Fig. 3.10-12
Jump resonance in a 2-phase network

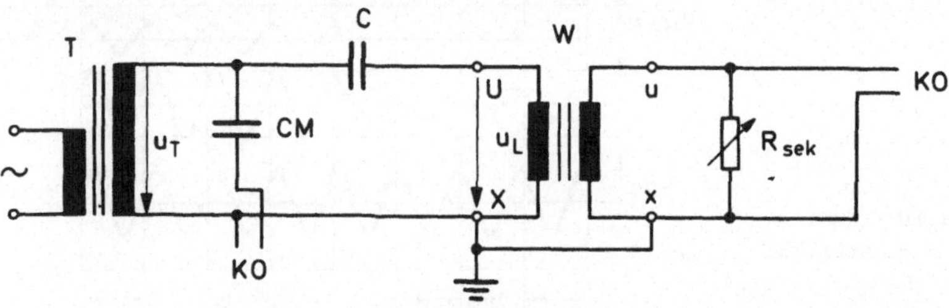

Fig. 3.10-13 Test setup for the measurement of jump resonance and subharmonic oscillations

In this circuit the transformer voltage and that of the voltage transformer should be determined shortly before and after jump resonance by gradually increasing the excitation. Finally that value of the loading resistor R_{sek}, connected in the secondary circuit of the voltage transformer, with which the jump resonance can be eliminated again, should be found.

e) Subharmonic Oscillations

To generate subharmonic oscillations the earth capacitance C of Fig. 3.10-4 must be substantially increased. For proper choice of C subharmonic oscillations can be induced by temporarily short-circuiting the secondary winding of the voltage transformer. The transition from the normal operating condition to the resonance state should be recorded on the storage oscilloscope for both cases. In addition it should be shown that the subharmonic oscillations can be prevented by adequate loading of the secondary winding of the voltage transformer with the resistance R_{sek}.

For this experiment at a transformer voltage of 17 kV and capacitance C = 13 000 pF, the 25 Hz subharmonic oscillations reproduced in Fig. 3.10-10 were obtained. In the series oscillatory circuit of Fig. 3.10-13 subharmonic oscillations of $16\frac{2}{3}$ Hz could be obtained for C = 7000 pF and U_T = 10 kV.

3.10.3 Evaluation

The characteristic $U_{eff} = f(I_{eff})$ for the voltage transformer should be drawn on graph paper. The jump resonance voltage U_k should be determined with the aid of the circuit data valid for experiment 3.10.2d using the graphical construction of Fig. 3.10-6, and compared with the measured value. In doing so an earth capacitance of 100 ... 300 pF should be taken into account, which is made up of the capacitances of the test transformer, voltage transformer and leads.

The current-voltage ellipse according to Fig. 3.10-8 should be calculated for the value of the resistance R_{sek} determined under 3.10.2d, drawn into the diagram of a) and briefly discussed.

Example: The determination of the jump resonance voltage U_k is carried out as represented by Fig. 3.10-6. For this purpose, the single-phase test setup of Fig. 3.10-13 is converted into the equivalent circuit of Fig. 3.10-5, applying Thevenin's theory:

From the requirement of equal capacitances, we have:

$$C = C_{11} + C_0$$

and from the requirement of equal open-circuit voltages,

$$U = \frac{C_0}{C_0 + C_{11}} U_T.$$

Using this data the slope of the straight line $(1/\omega C) I_{eff}$ can be calculated. The point of intersection of the straight line through A_k with the ordinate gives the jump resonance voltage U_k of the equivalent circuit, which, when multiplied by the factor $\frac{C_{11} + C_0}{C_0}$, represents the value to be compared with the experimentally determined jump resonance voltage.

The construction of the current-voltage ellipse should also be undertaken along the lines of Fig. 3.10-8 using the conversion method described above.

With the aid of the potential curves in Fig. 3.10-10 the stationary potential curves should be constructed for superimposed second subharmonic oscillation at the voltage transformers and at the midpoint capacitances.

Literature: *Rüdenberg* 1953; *Roth* 1959; *Lesch* 1959; *Sirotinski* 1966

3.11 Experiment "Travelling Waves"

A time-dependent variation in the electrical conditions at any point of a spatially extended system is registered by the other parts of the system in the form of electromagnetic travelling waves. If this change of state occurs in a time of the order of the transit times, the finite propagation velocity must be allowed for. This is always true for networks for energy transmission with long lines when the external or internal overvoltages occur with voltage varia-

tions in the range of microseconds to milliseconds. In laboratory practice with extremely large current and voltage variations in the nanosecond range, it is often necessary to consider the spatial setup of the circuit and the equipment, even when spread out over only a few meters, from the point of view of travelling wave theory.
The topics investigated in this experiment by measurements on low-voltage models fall under the following headings:

Lightning overvoltages
Switching overvoltages
Surge diverter
Protective range
Waves in windings
Impulse voltage distribution

It is assumed the reader has some basic knowledge of the propagation of electromagnetic waves on transmission lines.

3.11.1 Fundamentals

The differential element of a loss-free homogeneous transmission line can be described by its inductance L' and its capacitance C' per unit length. If $u(x, t)$ and $i(x, t)$ represent the voltage and current respectively at point x at time t, the solutions of the differential equations are [e.g. *Unger* 1967]:

$$u(x,t) = u_v(x - vt) + u_r(x + vt)$$
$$Z\,i(x,t) = u_v(x - vt) - u_r(x + vt)$$

Here

$Z = \sqrt{\dfrac{L'}{C'}}$, is the surge impedance and

$v = \sqrt{\dfrac{1}{L'C'}}$, the propagation velocity.

u_v and u_r are travelling waves which travel in the positive or negative x direction with velocity v; their time dependence for loss-free lines is determined by the initial or boundary conditions.
The maximum value v can reach is the velocity of light c. As a guide we have the following values:

for overhead lines $v \approx 300\ \mathrm{m}/\mu\mathrm{s} \approx c$; $Z \approx 500\ \Omega$
for underground cables $v \approx 150\ \mathrm{m}/\mu\mathrm{s}$; $Z \approx 50\ \Omega$

Adding the above solutions gives:

$$u = 2\,u_v - Z\,i.$$

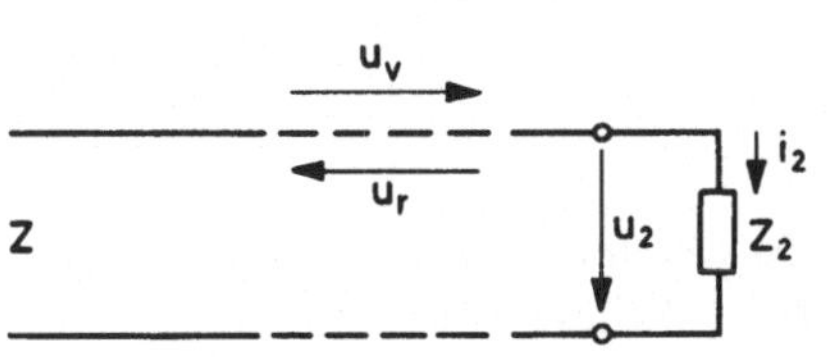

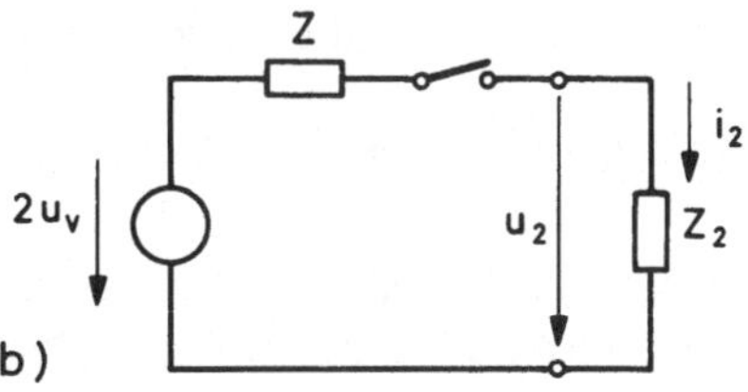

Fig. 3.11-1 Homogeneous transmission line terminated by Z_2
a) Circuit diagram b) Travelling wave equivalent circuit

This equation should be used to calculate the voltage u_2 at a point of reflection with input impedance Z_2 at the end of a homogeneous transmission line (Fig. 3.11-1a). For $u = u_2$ and $i = i_2$, an expression results which is reproduced by the travelling wave equivalent circuit of Fig. 3.11-1b.

The switch should be closed on arrival of the wave at the point of reflection. If τ_{min} is the lowest transit time occurring in a system for reflections travelling back to the point of reflection, the travelling wave equivalent circuit is valid for times $t \leqslant 2\tau_{min}$.

Introducing the reflection factor $r = u_r/u_v$, we have:

$$u_2 = u_v + u_r = u_v(1+r) = \frac{Z_2}{Z + Z_2}\, 2u_v$$

$$r = \frac{Z_2 - Z}{Z_2 + Z} \, .$$

The travelling wave equivalent circuit is particularly suitable for the determination of current and voltage at the end of an electrically "long" transmission line, or a line matched at the generator end.

a) Origin of Travelling Waves

Travelling waves as a consequence of lightning discharges

The front-times of the resulting travelling waves lie in the μs range, the tail-times are of the order of 100 μs. For a direct stroke to the conductor the transmission line is suddenly connected to a strong energy source. One may assume that a lightning current i_B is impressed and increases at rates of between 10 and 20 kA/μs. As a result of the lightning currents flowing in, current and voltage waves travel from the point of impact along the conductor.

For the case represented in Fig. 3.11-2 the voltage appearing at the point of impact is:

$$u = \frac{1}{2} Z i_B .$$

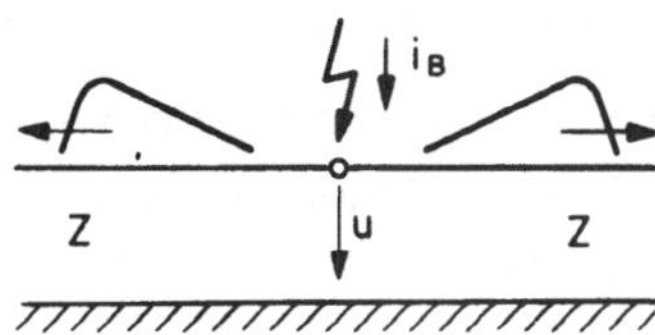

Fig. 3.11-2
Origin of travelling waves due to a lightning stroke

For overhead lines the rate of rise of the resulting overvoltage can be calculated as:

$$S = \frac{du}{dt} = \frac{1}{2} Z \frac{di_B}{dt} = 2.5 \dots 5 \text{ MV}/\mu s.$$

50 % of all the lightning strokes reach a peak value > 20 kA, and only 10 % of all the strokes have a maximum current > 60 kA. The amplitude of these lightning impulse voltages is limited to a value corresponding to the insulation level by flashovers at the insulator chains of the neighbouring masts. This value lies in the region of 2 to 5 times the peak value of the operating voltage.

For a direct stroke to the earth cable or to the mast of an overhead line, as a result of the earthing resistance, the mast can temporarily be at such a high potential that a reverse flashover occurs from the mast to one of the conductors. Hence the risk of a reverse flash-over occurring is especially large for unfavourable earthing conditions.

Moreover, external overvoltages can also be caused by an indirect stroke. Here the thunder cloud discharges itself in the vicinity of an overhead line in the form of a flash of lightning. The charge, induced on the conductor before the discharge, progresses along the line after the lightning discharge in the form of travelling waves. The amplitudes of waves following indirect strokes are comparatively low (up to about 100 kV), but are still dangerous for low-voltage networks and telephone systems.

Travelling waves as a consequence of switching operations

Internal overvoltages which occur as a result of switching operations are of special significance in ultra high-voltage networks. The amplitudes of these switching impulse voltages are only about 2 to 3 times the peak value of the operating voltage. However, since the electrical strength of inhomogeneous electrode configurations for large spacings in air is very low for switching impulse voltages, these largely define the dimensions of air clearances at high nominal voltages (⩾ 400 kV). The front-times are in the region of a few hundred μs and the tail-times in the ms range.

Particularly high overvoltages can occur during switching operations in connection with short-circuits or earth faults, as well as on switching-off unloaded transformers and capacitances (unloaded cables and overhead lines, capacitor batteries, etc.).

Travelling waves in laboratory work and testing practice

Extremely steep current and voltage variations quite often occur during breakdown mechanisms. Travelling waves are then induced on the conductors and in the measuring cables, which can lead to disturbances during measurement and endanger parts of the equipment. Travelling waves also occur on stressing electrical equipment with steep impulse voltages. The potential distribution in spatially extended insulation systems is affected by travelling waves. These phenomena also play a role in high-voltage generators which employ reflection mechanisms to generate high-voltage pulses (see section 1.3.4).

b) Limitation of Overvoltages Using Surge Diverters

The voltage at an installation can be limited with the aid of surge diverters. For high voltages the surge diverters are built up using a series multiple-gap section F and a current-

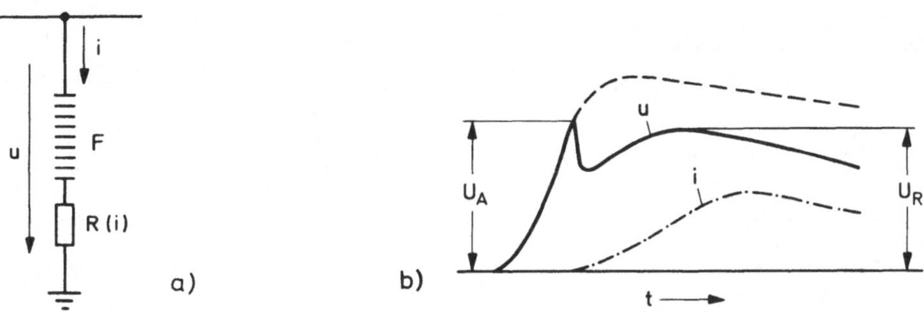

Fig. 3.11-3 Limitation of overvoltages by surge diverters
a) Equivalent circuit of a surge diverter
b) Time-dependent curves of voltage and current at the surge diverter

dependent resistance R(i), as shown in Fig. 3.11-3a. Should the terminal voltage be greater than the breakdown voltage U_A of the gap, it is reduced to the voltage drop across the resistance $U_R = iR(i)$. The terminal voltage u of the surge diverter and the current i for the limitation of an impulse voltage are shown in Fig. 3.11-3b; for the current curve i the effect of stray inductances has been taken into account. A surge diverter can guarantee reliable limitation of the voltage to U_A in every case at its terminals only. At a certain distance away from the diverter higher voltages may occur. The length of the conductor in front of or behind the surge diverter, within which a definite permissible overvoltage U_{zul} shall not be exceeded for a given waveform, is known as the protective range a. For square waves the full amplitude is present up to the surge diverter itself; reliable protection against travelling waves from either side is therefore effected only by setting up two surge diverters. The stretch of conductor lying in-between is then fully protected.

In practice one may assume that travelling waves have a finite voltage rate of rise S. The protective range for these wedge-shaped waves will be derived on the basis of Fig. 3.11-4. Let a surge diverter be installed at the point 2 along a homogeneous transmission line; at $t = 0$, the crest of the wave arrives at this position. The distance a between the points 1 and 2 is covered by the wave in time $\tau = a/v$. At $t = U_A/S$ the surge diverter responds to U_A and a backward wave is initiated, the shape of which can, for example, be determined with the aid of the travelling wave equivalent circuit. Assuming a linear increase of the incoming wave and ideal behaviour of the surge diverter, the reflection at point 2 is a wedge-shaped wave of slope $-S$. Only after the further interval of the time τ does a voltage limiting effect set in at the point 1, at the time $t = U_A/S + \tau$. At this instant the voltage u_1 has a value:

$$u_1 = U_A + 2S\tau.$$

Since u_1 is to be $\leqslant U_{zul}$, for the protective range we have:

$$a = \frac{U_{zul} - U_A}{2S} v.$$

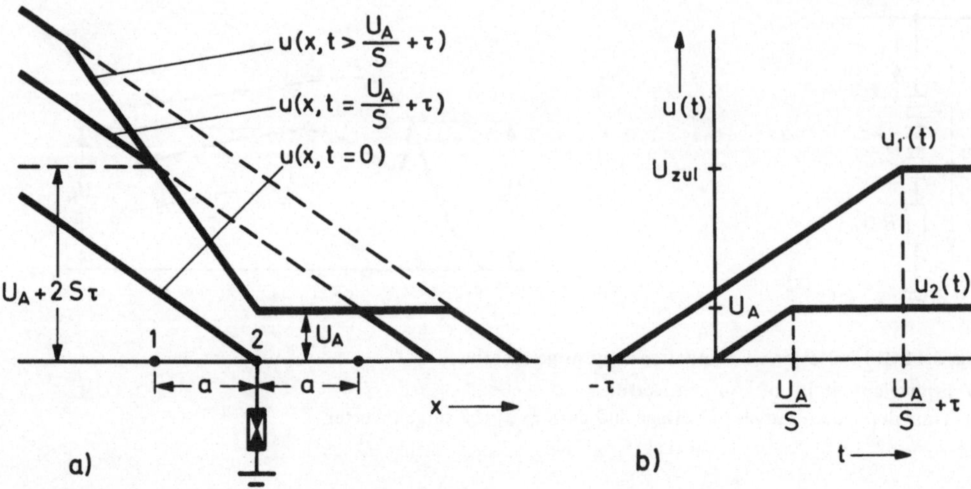

Fig. 3.11-4 Determination of the protective range of a surge diverter on impact of a wedge-shaped wave
a) u = f (x) b) u = f (t)

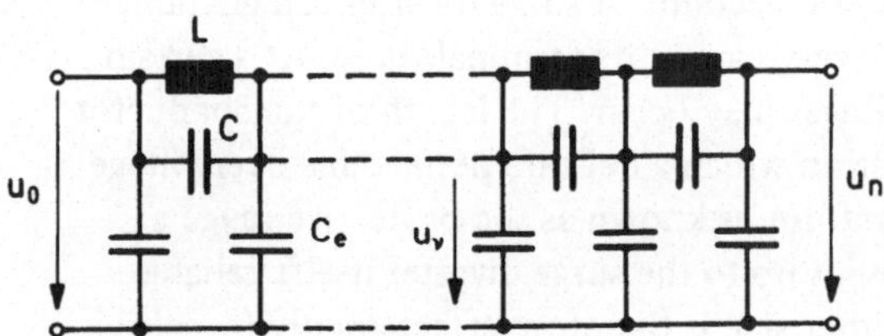

Fig. 3.11-5
Equivalent circuit of a transformer winding for impulse voltage stress

c) Travelling Waves in Transformer Windings

Development of an equivalent circuit

A technically important special case of travelling wave propagation is the impact of an impulse voltage in a transformer winding. Whereas at low frequencies the voltage distribution in the winding is linear because of the magnetic flux linking all the turns, at higher frequencies, as in the spectrum of an impulse voltage, it is also determined by the capacitances. To a first approximation one can perform the calculations using the equivalent circuit of a high-voltage winding (Fig. 3.11-5), suggested by *K. W. Wagner* in 1915. L and C denote the inductance and capacitance respectively of an element of the winding and C_e the corresponding earth capacitance.

An impulse voltage $u_0(t)$ applied to the high-voltage terminals will travel with a finite velocity along the individual turns to the end of the winding, be reflected there, and so on. Superimposed on this process however, is a wave which takes a shorter path, namely primarily via the capacitive coupling between individual parts of the winding. The oscillations produced inside the winding in this way result in a time-varying, asymmetrical voltage distribution, which may cause individual parts of the winding to be gravely overstressed.

Calculation and control of the voltage distribution

At the instant of impact of a wave (t = 0) the capacitances alone determine the voltage distribution in the winding. Hence, for the calculation of the initial distribution for short waves $L = \infty$ can be assumed. It can be shown that for the simple case of a square wave of amplitude U_0 on the earthed winding ($u_n = 0$) as in Fig. 3.11-5, the following relationship holds:

$$U_\nu = U_0 \frac{\sinh(n-\nu)\alpha}{\sinh n\alpha}, \quad \text{with} \quad \alpha = \sqrt{\frac{C_e}{C}} .$$

The part of the capacitive equivalent circuit shown in full lines in Fig. 3.11-6a corresponds to the requirements of the voltage distribution indicated by (C, C_e) in Fig. 3.11-6b.

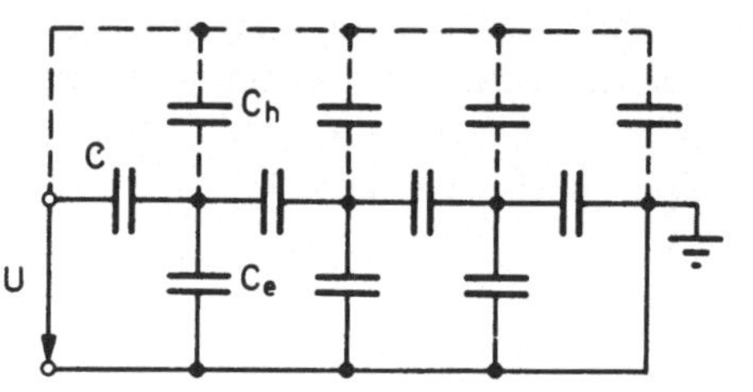

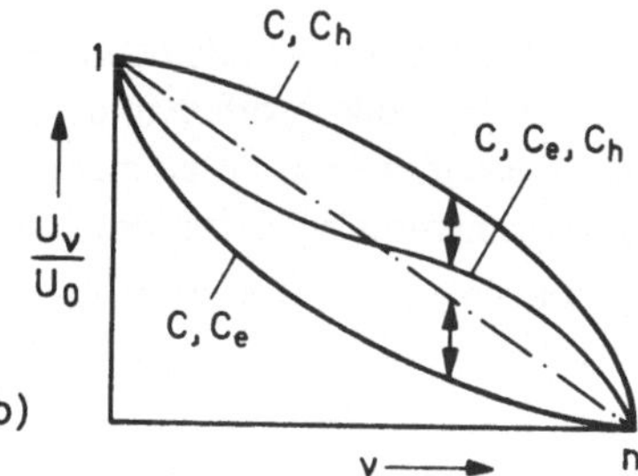

Fig. 3.11-6 Impulse voltage distribution in a transformer winding
a) Equivalent circuit with effective interturn and coupling capacitances
b) Voltage distribution on impact of a wave

There is also quite often capacitive coupling to the high-voltage electrode besides that to earth; this is taken into account by the elements C_h in Fig. 3.11-6a.

The capacitive initial voltage distribution, taking account of C_e and C_h at the same time, can no longer be calculated in a straightforward way. For not too great a divergence from the linear distribution the resultant distribution can be approximately derived from Fig. 3.11-6b, whereby, for a given ν, the amount of deviation from the linear distribution produced by C_e only is deducted from that value effected by C_h alone [*Philippow* 1966].

The capacitances can be varied by appropriate construction of the transformer, and the voltage distribution thereby influenced to advantage. An increase of C relative to C_e effects linearisation. This may be achieved by the chosen type of winding, for example by layer winding or by disc winding. Another possibility is the installation of large-area electrodes (shields) at the high-voltage terminals of the winding; these increase the coupling capacitances C_h. The current taken by the earth capacitances then flows partly through the high-voltage capacitances C_h. A linear voltage distribution is obtained in the transformer when for each winding element the current flowing across C_h is equal to that flowing across C_e.

The smaller the divergence of the initial distribution from the linear final distribution, stressing of the winding in the voltage rise would not only be more uniform, but the amplitudes of the subsequent transient phenomena would also be reduced. Windings with a linear initial distribution are often characterized as non-oscillatory for this reason. Particularly severe stressing occurs during a test with chopped impulse voltages because of the rapid voltage variations which then take place.

3.11.2 Experiment

a) Model Investigations at Low Voltage

Travelling wave investigations of networks, and in particular of transformer windings, are often carried out on models. For this purpose a generator is commonly used which can supply impulse voltages with adjustable waveform and with a peak voltage of a few hundred volts at a definite impulse repetition frequency of, say, 50 Hz. A repetitive surge generator RG[1]) of this kind is available for this experiment. By synchronising the generator and the oscilloscope, a stationary diagram can be obtained on the KO screen.

With these generators real power transformers can be impulse tested before they are mounted inside the tank in order to determine the impulse voltage distribution; or for design purposes measurements can be made on a model. In a similar manner it is also possible to simulate transmission lines, for which either cables or four-pole chains may be used.

For convenient realization of the transmission line model, a coaxial delay cable with $Z = 1500\ \Omega$ and $v = 4.0\ m/\mu s$ was chosen for this experiment.

A circuit with a Zener diode is suitable as a model of a surge diverter. The Zener voltage of about 30 V corresponds to the residual voltage of the diverter.

A double-beam cathode-ray oscilloscope with a bandwith $\geqslant 10$ MHz should be available for the measurements.

b) Travelling Waves on Transmission Lines

An overhead line should be reproduced as in Fig. 3.11-7 by two cables K1 and K2, each with a transit time of $\tau/2$.

The beginning of the line is terminated by the surge impedance of the cable, so that the output terminals $0 - 0'$ of the connected repetitive surge generator RG can be considered to be short-circuited by a large capacitance. The generator should be adjusted to deliver a 1.2/50 impulse voltage with connected load. The voltage curves at the terminals $1 - 1'$, $2 - 2'$ and $3 - 3'$ should be oscillographed for various operating conditions. The measurement should be performed for various terminations of the transmission line at the terminals $3 - 3'$. As an example, the curves for an open-circuited line ($Z_3 = \infty$) are reproduced in Fig. 3.11-8, and for a line terminated by a capacitance in Fig. 3.11-9.

c) Protection by Surge Diverters

The model of the surge diverter should be connected at position $2 - 2'$ in the circuit of Fig. 3.11-7. The voltage curves should be oscillographed at the terminals $1 - 1'$, $2 - 2'$ and $3 - 3'$, the transmission line being open-circuited ($Z_3 \to \infty$). The curves are reproduced in Fig. 3.11-10 as an example.

---

[1]) Manufacturer: Emil Haefely & Cie AG, Basel

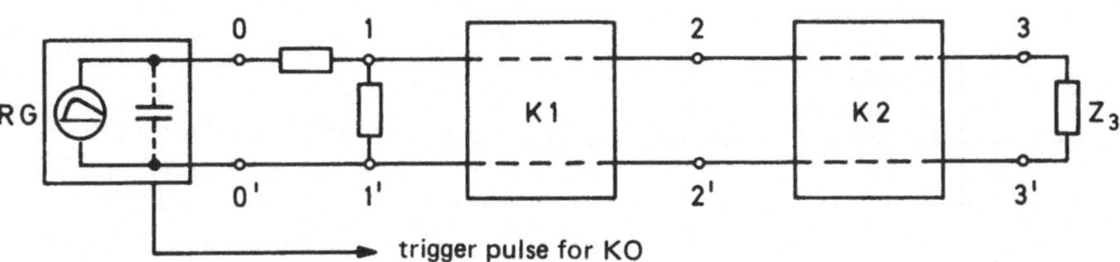

Fig. 3.11-7 Low-voltage model for travelling wave investigations on transmission lines
RG: Repetitive surge generator, K1, K2: Cables with v = 4 m/μs, each 12 m long

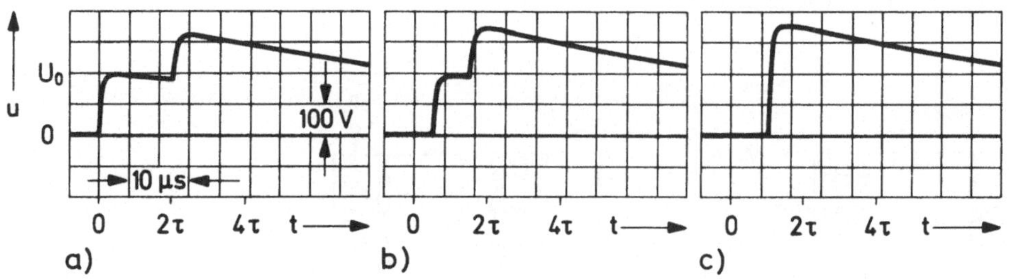

Fig. 3.11-8 Oscillograms of voltage curves for the transmission line model in Fig. 3.11-7 for $Z_3 \to \infty$
a) Measurement at the terminals 1 - 1′ b) Measurement at the terminals 2 - 2′
c) Measurement at the terminals 3 - 3′

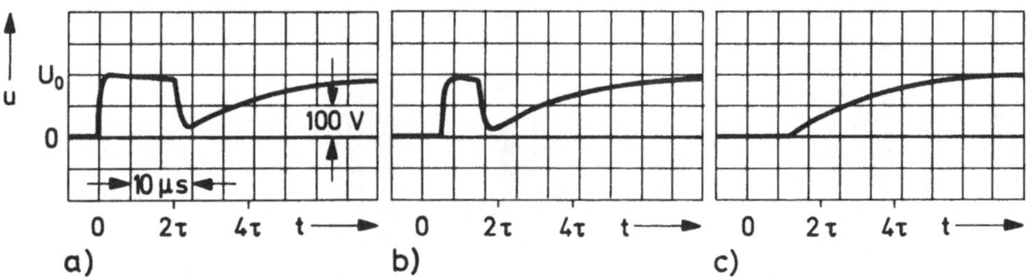

Fig. 3.11-9 Oscillograms of the voltage curves for Z_3 = C
a) Measurement at the terminals 1 - 1′ b) Measurement at the terminals 2 - 2′
c) Measurement at the terminals 3 - 3′

d) Impulse Voltage Distribution in Transformer Windings

For these investigations a transformer model was rebuilt with interchangeable high-voltage winding. Thus high-voltage windings with the same number of turns and external dimensions, but with very different properties, can be investigated. In the one case the high-

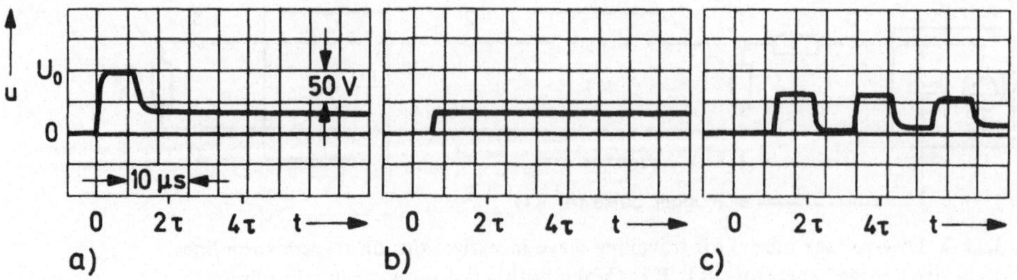

Fig. 3.11-10 Oscillograms of the voltage curves for the installation of a surge diverter at position 2 - 2′ ($Z_3 \rightarrow \infty$)

a) Measurement at the terminals 1 - 1′ b) Measurement at the terminals 2 - 2′
c) Measurement at the terminals 3 - 3′

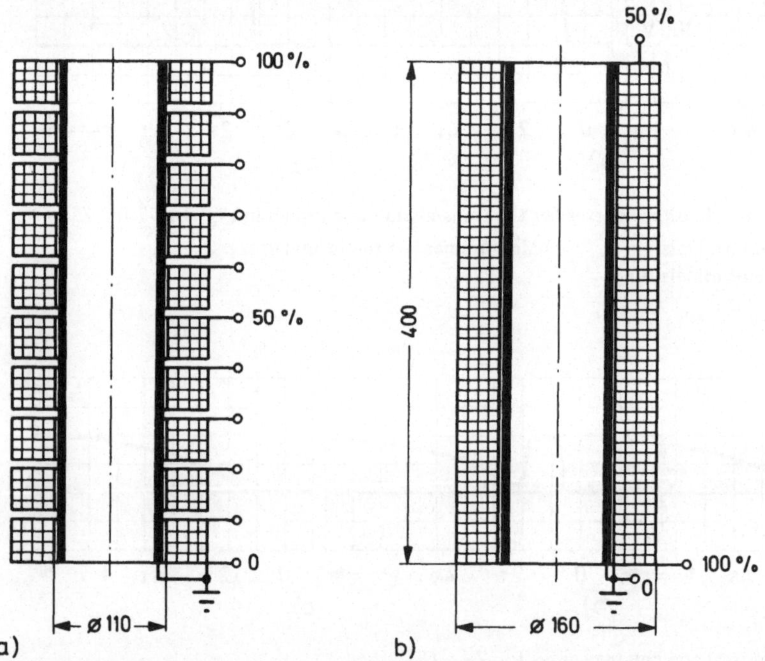

Fig. 3.11-11 Transformer model for travelling wave investigations

a) High-voltage winding in the form of disc winding
b) High-voltage winding in the form of layer winding

voltage winding consists of disc winding (Fig. 3.11-11a) and in the other case layer winding (Fig. 3.11-11b). Both windings are provided with measuring taps at every 10 % of the total number of turns (= 1800). The low-voltage winding is connected to earth.

The repetitive surge generator RG should be directly connected to the transformer model. Voltage curves at the various taps should be oscillographed for different cases:

I High-voltage winding as disc winding (Fig. 3.11-12)
II High-voltage winding as layer winding (Fig. 3.11-13).

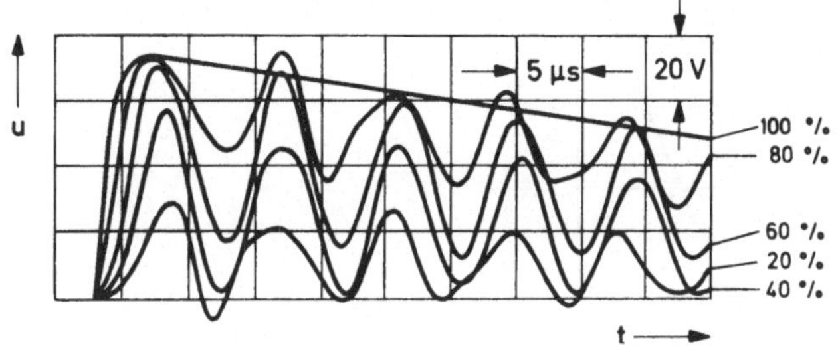

Fig. 3.11-12
Voltage curves of the transformer model with disc winding during impulse voltage stress

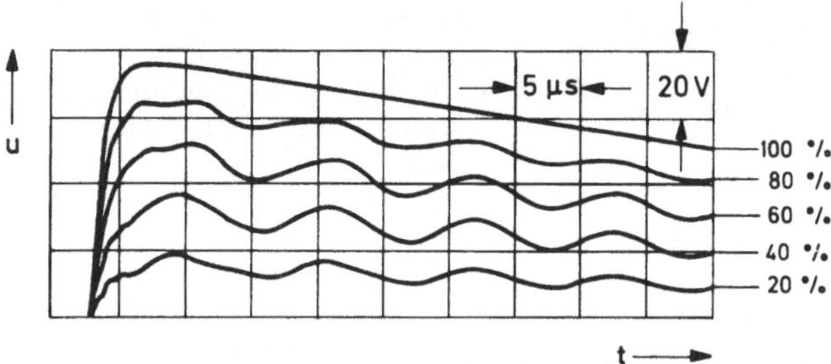

Fig. 3.11-13
Voltage curves of the transformer model with layer winding during impulse voltage stress

3.11.3 Evaluation

The lengths of a real transmission line or a real underground cable should be determined corresponding to the functional model investigated in section 3.11.2b. How large is the ratio of the amplitude of the voltage generated by the repetitive surge generator RG to that of the voltage entering the line at $1-1'$?

For the cases investigated under 3.11.2b the idealised curves for a rectangular shaped voltage and an ideal cable should be drawn, and compared with the recorded oscillograms.

For the oscillograms of the measurement according to 3.11.2c with the diverter model, idealised curves should be shown and compared with the measured results. The protective range a should be calculated for an assumed permissible voltage U_{zul} = 40 V at the line end.

From the oscillograms of 3.11.2d, for the cases I and II, the position, time and amplitude should be determined of the highest voltage stress between two measuring points.

For case I the voltage distribution in the transformer winding $u = f(\nu)$ (with ν in %) should be graphically plotted for times $t = 1.5\ \mu s$ and $t = 15\ \mu s$ (impact of the wave at time $t = 0$) (Fig. 3.11-14). Where is the maximum stress in the winding for these times?

Literature: *Bewley* 1951; *Strigel* 1955; *Rüdenberg* 1962; *Sirotinski* 1965; *Philippow* 1966; *Heller, Veverka* 1968; *Greenwood* 1971

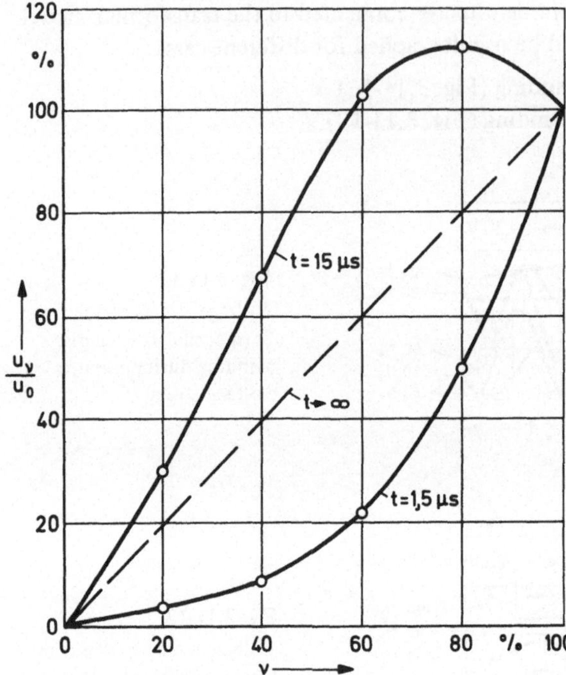

Fig. 3.11-14
Voltage distribution in a transformer model with disc winding at various times t after impact of a travelling wave

3.12 Experiment "Impulse Currents and Arcs"

Transient high currents are important to numerous fields of science and technology, because of their magnetic, mechanical and thermal effects. Whenever strong currents coexist with an arc, a heavy concentration of energy results which can also lead to destruction. Lightning currents or the short-circuit currents in high-voltage networks are examples for this. The topics covered by this experiment fall under the following headings:

Discharge circuit with capacitive energy storage
Impulse current measurement
Forces in a magnetic field
Alternating current arc
Arc quenching

It is assumed the reader is familiar with section 1.4 Generation and measurement of impulse currents.

3.12.1 Fundamentals

a) Force Action in a Magnetic Field

The force action in a magnetic field should be demonstrated using the transformer arrangement of Fig. 3.12-1. Position and dimensions of the outer winding 1 carrying a current $i_1(t)$ should be invariable, so that only the forces effective on the shorted inner winding 2 are to be observed. Radial forces $\vec{F}_r$ tend to reduce the diameter, and the axial forces $\vec{F}_a$ the length of the inner winding. In a fully symmetrical arrangement (a = 0), all the forces acting at the centre of gravity of the inner winding just compensate each other and the system is in a state of unstable equilibrium. On the other hand, if the inner winding is axially displaced ($a \neq 0$), a resultant axial force $\vec{F}$ is produced tending to increase the asymmetry.

The inductance L of the arrangement, measured at the terminals of the outer winding, increases with a from the value for a = 0 to the value for the completely removed inner winding. The magnetic energy

$$W = \frac{1}{2} L i_1^2$$

also varies accordingly.

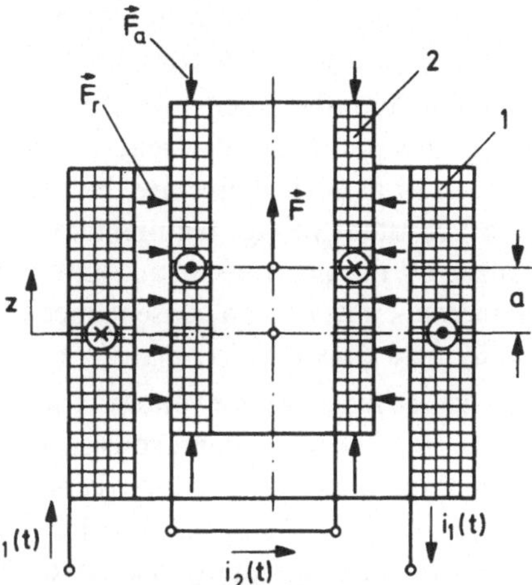

Fig. 3.12-1
Transformer arrangement
1 Primary winding
2 Secondary winding

For the resultant force acting at the centre of gravity of the inner winding we have:

$$F(z) = \frac{dW}{dz} = \frac{1}{2} i_1^2 \frac{dL}{dz} + \frac{1}{2} L \cdot \frac{d(i_1^2)}{dz} .$$

The current pulse is often of such short duration that the reaction of the motion does not affect the magnitude of the force. In this case, since z = a, the inductance of the arrangement remains constant during the current interaction period. The initial velocity v_0 of the inner winding with mass m can then be calculated as follows, according to the law of conservation of momentum:

$$v_{0(z=a)} = \frac{1}{m} \int_0^\infty F\,dt = \frac{1}{2m} \left[\left(\frac{dL}{dz}\right)_{(z=a)} \cdot \int_0^\infty i_1^2\,dt + L \cdot \frac{d}{dz} \left(\int_0^\infty i_1^2\,dt \right)_{(z=a)} \right].$$

The kinetic energy supplied to the inner winding by the current pulse is then:

$$W_{kin} = \frac{1}{2} m v_0^2 .$$

The reliable control of the mechanical forces in electrical circuits among other things determines the short-circuit strength of transformers. These forces can also be made use of in metal-forming, where the workpiece to be formed takes the place of the shorted inner winding of the transformer model. A further useful application is in electrodynamic drives. Here those arrangements are particularly suitable in which the variation of inductance due to movement of the object to be accelerated is as large as possible. Even for the case of the transformer arrangement discussed above, a driving action may be observed if the inner winding is replaced by a movable conductor, e.g. a metal tube.

b) Alternating Current Arc

Arcs constitute a type of gas discharge very important to electric power technology (see 3.7.1c). They occur, for example, in switching devices on current switch-off, thereby preventing an abrupt interruption of the latter, and with that dangerous over-voltages.

As indicated in the schematic representation of Fig. 3.12-2, arcs possess two space-charge regions in front of the electrodes, which are described as the anode drop and the cathode drop. In air at normal pressure their length is of the order of 1 μm and both have an almost constant voltage requirement of about 10 V each. Current transport is essentially realized by the electrons on account of their high mobility. They are created at the cathode by thermo-emission as a consequence of heating of the cathode base by bombarding ions and by field emission.

At large arc lengths s the voltage requirement of the arc column determines the total arc voltage u_b. The arc column consists of plasma in which the charge carrier loss, due to recombination and dissipation, is compensated by thermo-ionisation. The column temperature lies in the region of 6000 to 12 000 K. Consequently arcs can only exist when an adequately high energy supply is assured. Stationary arcs require a minimum current of about 1 A and are ignited either by contact breaking or by a breakdown process. If the current supplied by the energy source is insufficient for the existence of an arc, either a glow discharge of appreciably lower current intensity results, or total current interruption.

The power fed into the arc is given by:

$$P_{zu} = i_b u_b .$$

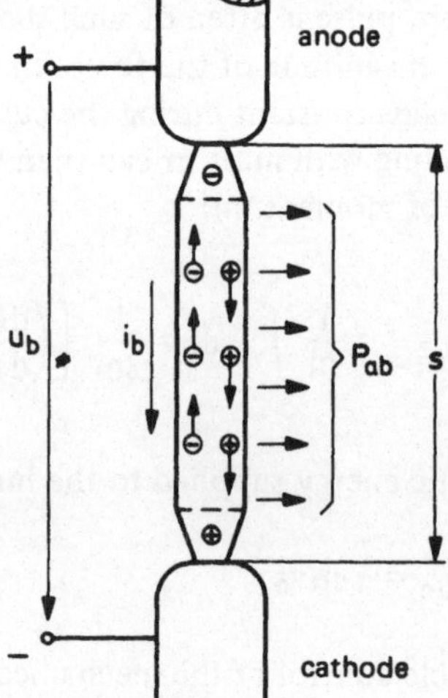

Fig. 3.12-2
Schematic representation of an arc

The heat extraction P_{ab} takes place for short arcs via the electrodes, otherwise mainly by thermal conduction, convection and radiation out of the arc column. This can be influenced by the electrode material (carbon, metal) and by the coolant of the arc column (air, oil). If the heat quantity Q stored in the arc is also taken into account, for power equilibrium we have:

$$i_b u_b - P_{ab} = \frac{dQ}{dt} .$$

Neglecting the thermal capacitance (Q = 0), we obtain the following equation for the arc characteristic:

$$u_b = \frac{P_{ab}}{i_b} .$$

The simple assumption P_{ab} = const. is useful in many cases and results in a hyperbolic dependence of the arc voltage upon the current. Cooling of the arc may also result in heat extraction which is approximately proportional to the current; the arc voltage then remains fairly constant. This is the case, for example, for very short spacings between metal electrodes, since the energy converted in the drop-zones with constant voltage requirement is conducted away through the electrodes.

The arc voltages for both the limiting cases P_{ab} = const. and $P_{ab} \sim i_b$ are shown dotted in Fig. 3.12-3 for the case of a sinusoidal current. A stationary alternating current arc must be re-ignited after each crossover of the current. Moreover, the thermal energy stored in the arc may not be neglected. The arc voltage requirement therefore corresponds rather more to the curves drawn in full lines. The voltage value U_Z is known as the ignition peak and that of U_L as the quenching peak.

c) Arc Quenching

Whether or not an alternating current arc re-ignites after a current zero depends upon the properties of the arc column and the circuit. The speed of de-ionisation of the arc path,

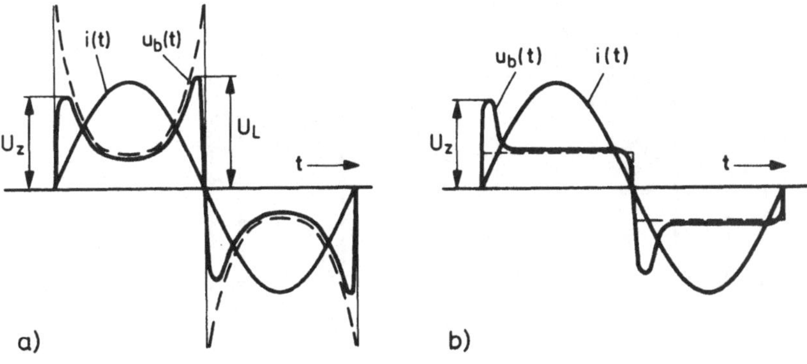

Fig. 3.12-3 Time variation of the current and voltage of an arc
a) P_{ab} = const. b) $P_{ab} \sim i_b(t)$

and the transient recovery voltage, which develops along the arc path on interruption of the current, are both critical. The de-ionisation process can essentially be influenced by the cooling of the arc. The form of the transient recovery voltage is determined experimentally, or calculated by neglecting the effect of the arc [*Slamecka* 1966].

Reliable interruption of a.c. arcs other than in switchgear should also be ensured for the gaps of surge diverters. Here however, in contrast to switchgear, the arc is ignited not by the interruption of a metallic current path but by a breakdown caused by exceeding the breakdown voltage. The a.c. arc so produced is fed from the network mains and must be reliably interrupted by the gap, since otherwise the diverter would be overloaded and destroyed.

For effective cooling of the arc, diverter gaps are built up in a series circuit of several gaps with low spacing using electrodes of large area with high cooling efficiency. Particularly good quenching properties are obtained when the arc is additionally cooled by magnetic deflection.

3.12.2 Experiment

a) Measurement of Current in the Discharge Circuit with Capacitive Energy Storage

The experimental circuit is shown in Fig. 3.12-4. An impulse capacitor C of capacitance 7.5 μF is charged via a charging resistance R_L to a direct voltage U_0, which can be measured across the measuring resistance R_m. The heavily drawn discharge circuit should be as low-inductive as possible, for example using strip conductors. The trigger gap F can be set up as in Fig. 2.4-7 and is controlled by a trigger generator ZG (see Fig. 2.4-6) via a transformer ZT, intended for potential separation. The current i flowing through the test arrangement P should be measured across the measuring resistance R_i.

Fig. 3.12-5 shows a coaxially arranged measuring resistor for current measurements in the path of the strip conductor 1; its flat copper conductors are insulated against each other by plastic foils. The current flows through the mounting ring 2 and the screening chamber 3 to the resistance element 4, arranged between the contact discs 5 and 6. It returns to the strip conductor via the threaded connector 7 of the upper contact disc. The two contact

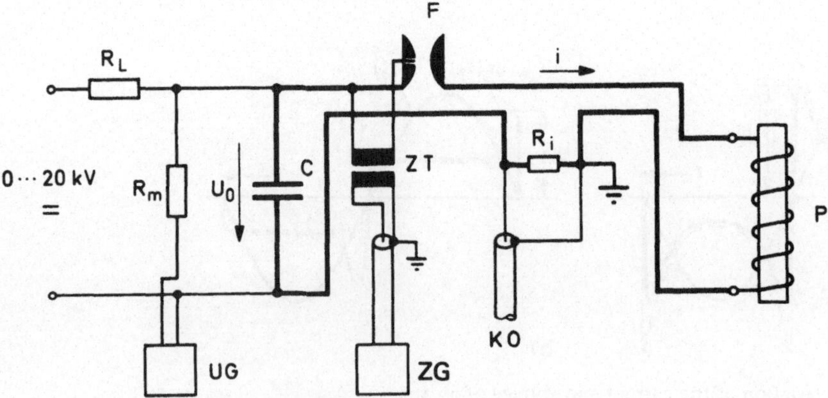

Fig. 3.12-4 Impulse current circuit with capacitive energy storage

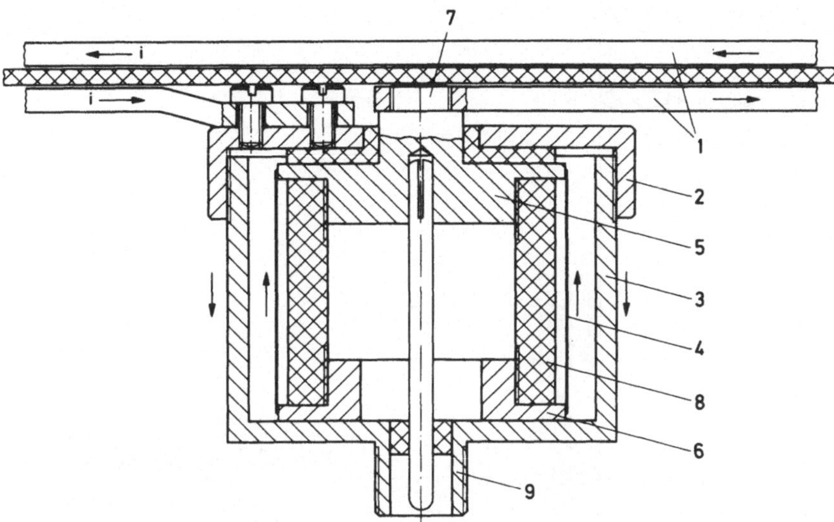

Fig. 3.12-5 Coaxial measuring resistance in the path of a strip conductor

1 Strip conductor
2 Fixing ring
3 Screening chamber
4 Resistance element
5 and 6 Contact discs
7 Threaded connector
8 Insulating distance ring
9 Connector terminal

discs are separated by the insulating tube 8. The measuring voltage proportional to the current is tapped at the connector terminal 9. To keep the inductance of the measuring resistor at a low value, the spacing between the chamber 3 and the resistance element 4 should be as small as possible.

The testing coil shown in Fig. 3.12-6 should be investigated as the test setup. The discharge current is fed to the cylindrical primary winding 1, the inductance of which is about 26 μH in a model with the given dimensions and 25 turns. So that the winding is not destroyed by the forces during the experiment, it should be enclosed in an insulating material chamber comprising the parts 2 and 3. The metal tube 4 arranged coaxially with 1 functions as a shorted secondary winding. The tube can be axially displaced on the insulating body 5. Terminals 6 of the primary winding are connected to the strip conductor 8 with the aid of bracing 7.

The time dependence of the current i should be oscillographed for the following three cases, at a charging voltage U_0 = 20 kV:

Testing coil P replaced by short-circuiting link,
Secondary winding 2 of the testing coil in the position shown in Fig. 3.12-6 (a = 0),
Testing coil without secondary winding.

The three oscillograms reproduced in Fig. 3.12-7 were obtained for this experiment. Since the capacitance C is known, using the periodic time the total inductance of the circuit can

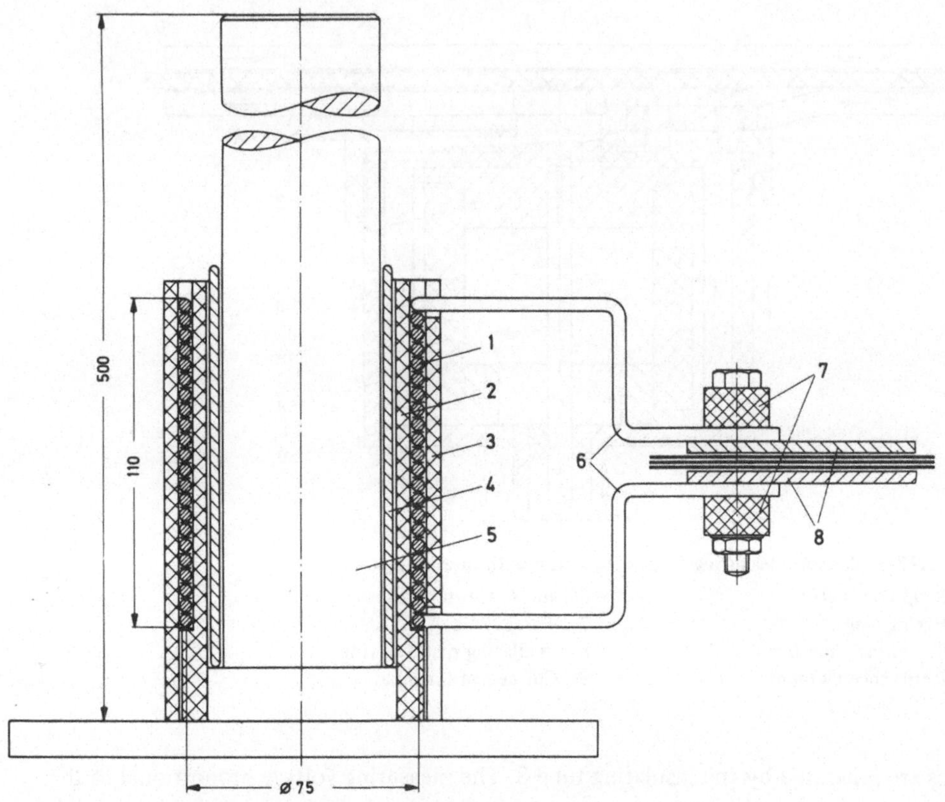

Fig. 3.12-6 Testing coil for impulse current investigations

1 Primary winding, 25 turns of 6 mm² copper
2 and 3 Insulating material chamber
4 Secondary winding
5 Insulating body
6 Connecting ends
7 Strip conductor bracing
8 Strip conductor

be determined and also, by subtraction, that of the individual circuit elements. The following values were obtained:

Inductance of the setup: 0.238 μH
Inductance of the experimental arrangement with secondary winding: 11.5 μH
Inductance of the experimental arrangement without secondary winding: 25.9 μH

b) Investigation of Electrodynamic Forces

In the same circuit the discharge current is allowed to flow through the experimental arrangement of Fig. 3.12-6 with its secondary winding displaced upwards by a = 6 cm using an insulating tube. The secondary winding then experiences an upward acceleration, which results in a certain lift H. For a suitably chosen charging voltage U_0 the lift obtained may be determined. At the same time the current oscillogram should be recorded.

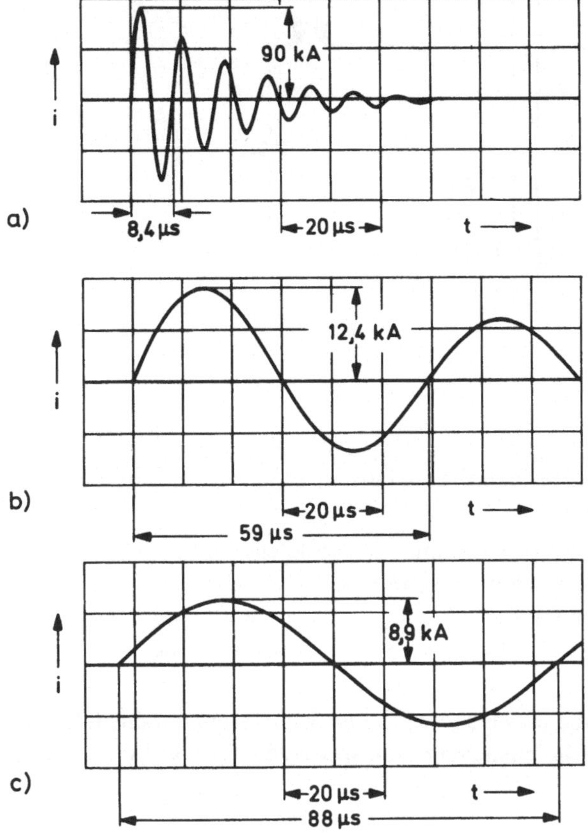

Fig. 3.12-7
Current oscillograms for determination of inductance
a) Test object short-circuited
b) Secondary winding in central position
c) Without secondary winding

For this experiment no frequency variation was obtained over the entire current curve. From this it follows that the local displacement of the secondary winding begins practically only after the current has already died away; the conditions for a pulse drive are therefore given. In the investigated arrangement a lift of H = 20 cm was obtained for a charging voltage U_0 = 16.5 kV. Here the mass of the accelerated tube was 283 g. By equating the potential and kinetic energies an initial velocity v_0 = 1.98 m/s can then be calculated.

c) The Stationary a.c. Arc

For this experiment the circuit of Fig. 3.12-8 should be set up without the parts shown in dotted lines. The a.c. arc should be fed from the low-voltage network. For reasons of safety, an isolating transformer T with ratio 220/220 V should be provided. The experimental setup consists of two cylindrical electrodes 1 and 2 of diameter d. The spacing s can be adjusted using the earthed electrode and an arc discharge struck by contact separation. The current can be adjusted with the inductance $L_K \approx 50$ mH and the resistance $R_K = 0 \ldots 20\ \Omega$, when values up to 10 A should be reached. The arc voltage u_b is measured

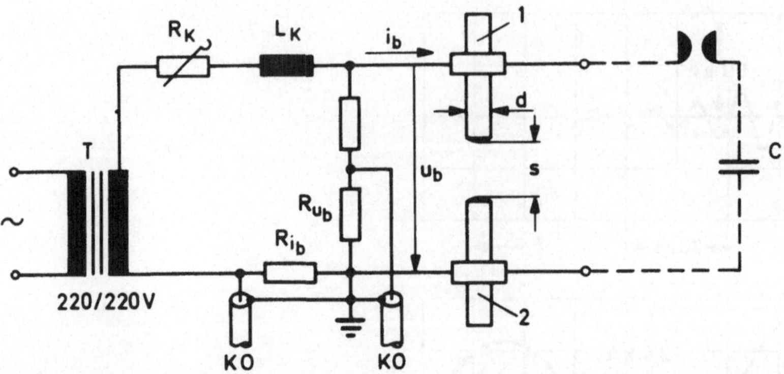

Fig. 3.12-8 Circuit for investigation of a.c. arcs. 1, 2 Electrodes

at the resistance R_{u_b} of a potential divider and the arc current i_b across a measuring resistance R_{i_b}.

Using carbon electrodes with d = 8 mm and s = 5 mm, u_b and i_b should be oscillographed as a function of time; the arc is struck by contact separation. By variable fanning of the arc with air and at the same time observing the curve of u_b, the effect of cooling on the arc voltage should be qualitatively observed. Further, the arc characteristic $u_b = f(i_b)$ should be oscillographed.

For this experiment in stationary air the curves shown in Fig. 3.12-9 were obtained. On fanning the arc, the ignition peak and the quenching peak appeared clearly at first, until finally the arc was extinguished.

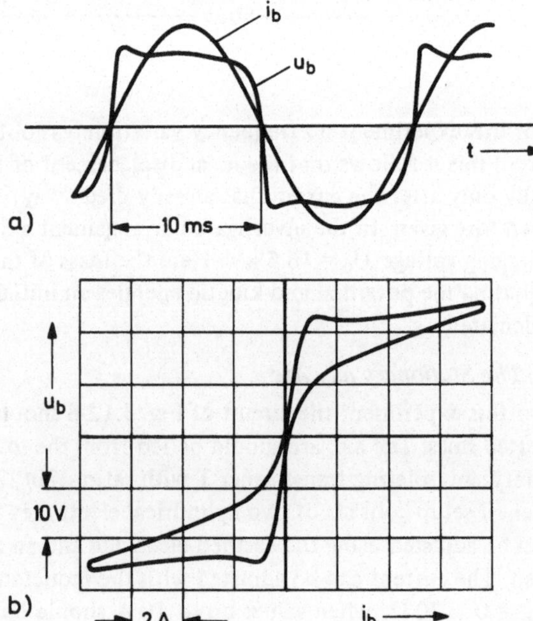

Fig. 3.12-9
Oscillograms of current and voltage of a stationary a.c. arc, carbon electrodes. s = 5 mm
a) Time dependence
b) Dynamic arc characteristic

d) Operation Test of a Diverter Gap

For these investigations the a.c. arc should be ignited by an impulse current discharge. That part of the circuit shown in dotted lines in Fig. 3.12-8 should be used for this purpose; it symbolizes the circuit for generating impulse currents according to Fig. 3.12-4. During these experiments the inductance L_K protects the transformer against overvoltages. Carbon electrodes should be used at first. The spacing s should be adjusted so that the breakdown voltage lies below the charging voltage of the capacitor C. The arc initiated by the discharge of the capacitor should be observed. The experiment should then be repeated with brass electrodes.

For this experiment reliable striking of the arc by the impulse current circuit was only assured with carbon electrodes. The arc current was about 10 A at a gap spacing of 4 mm. For brass electrodes the electrode spacing had to be reduced to about 0.2 mm in order to obtain a stationary arc at all.

3.12.3 Evaluation

The inductance of the testing coil, with and without the secondary winding, should be determined from the current oscillograms recorded in section 3.12.2a.

The efficiency of the mechanical motion achieved in the investigations under section 3.12.2d should be calculated. To this end, the ratio of the potential energy of the secondary winding to the capacitvely stored energy in the impulse capacitor should be derived.

The relationship $P_b = f(t)$ should be constructed from the curves of i_b and u_b recorded under section 3.12.2c.

The different quenching behaviour of carbon and brass electrode arrangements observed in section 3.12.2d should be discussed.

Appendix 1

Safety Regulations for High-Voltage Experiments

Experiments with high-voltages could become particularly hazardous for the participants should the safety precautions be inadequate. To give an idea of the required safety measures, as an example the safety regulations of the High-Voltage Institute of the Technical University of Braunschweig shall be described below. These supplement the appropriate safety regulations and as far as possible prevent risk to persons. Strict observance is therefore the duty of everyone working in the laboratory. Here any voltage greater than 250 V against earth is understood to be a high voltage.

Fundamental Rule: Before entering a high-voltage setup everyone must convince himself by personal observation that all the conductors which can assume high potential and lie in the contact zone are earthed, and that all the main leads are interrupted.

Fencing

All high-voltage setups must be protected against unintentional entry of the danger zone. This is appropriately done with the aid of metallic fences. When setting up the fences for voltages up to 1 MV the following minimum clearances to the components at high-voltage should not be reduced:

for alternating and direct voltages	50 cm for every 100 kV
for impulse voltages	20 cm for every 100 kV

A minimum clearance of 50 cm shall always be observed, independent of the value and type of voltage. For voltages over 1 MV, in particular for switching impulse voltages, the values quoted could be inadequate; special protective measures must then be introduced.

The fences should be reliably connected with one another conductively, earthed and provided with warning boards inscribed: "High-voltage! Caution! Highly dangerous!" It is forbidden to introduce conductive objects through the fence whilst the setup is in use.

Safety-Locking

In high-voltage setups each door must be provided with safety switches; these allow the door to be opened only when all the main leads to the test setup are interrupted.

Instead of direct interruption, the safety switches may also operate the no-voltage relay of a power circuit breaker, which, on opening the door, interrupts all the main leads to the setup. These power circuit breakers may only be switched on again when the door is closed. For direct supply from a high-voltage network (e.g. 10 kV city network), the main leads must be interrupted visibly before entry to the setup by an additional open isolating switch.

The switched condition of a setup must be indicated by a red lamp "Setup switched on" *and* by a green lamp "Setup switched off".

If the fence is interrupted for assembly and dismantling operations on the setup, or during large-scale modifications, all the prescribed precautions for entry of the setup shall be observed. Here particular attention must be paid to the reliable interruption of the main leads. On isolating switches or other disconnecting points, and on the control desk of the setup concerned, warning boards inscribed "Do not switch on! Danger!" must be displayed.

Earthing

A high-voltage setup may be entered only when all the parts which can assume high-voltage in the contact zone are earthed. Earthing may only be effected by a conductor earthed inside the fence. Fixing the earthing leads onto the parts to be earthed should be done with the aid of insulating rods. Earthing switches with a clearly visible operating position, are also permissible. In high-power setups with direct supply from the high-voltage network, earthing is achieved by earthing isolators. Earthing may only follow after switching the current source off, and may be removed only when there is no longer anyone present within the fence or if the setup is vacated after removal of earth. All metallic parts of the setup which do not carry potential during normal service must be earthed reliably and with an adequate cross-section of at least 1.5 mm^2 Cu. In test setups with direct supply from the high-voltage network, the earth connections must be made with particular consideration of the dynamic forces which can arise.

Circuit and Test Setup

Inasmuch as the setup is not supplied from ready wired desks, clearly marked isolating switches must be provided in all leads to the low-voltage circuits of high-voltage transformers and arranged at an easily identifiable position outside the fence. These must be opened before earthing and before entering the setup.

All leads must be laid so that there are no loosely hanging ends. Low-voltage leads which can assume high potentials during breakdown or flashovers and lead out of the fenced area, e.g. measuring cable, control cable, supply cable, must be laid inside the setup in earthed sleeving. All components of the setup must be either rigidly fixed or suspended so that they cannot topple during operation or be pulled down by the leads.

For all setups intended for research purposes, a circuit diagram shall be fixed outside the fence in a clearly visible position.

A test setup may be put into operation only after the circuit has been checked and permission to begin work given by an authorized person.

Conducting the Experiments

Everyone carrying out experiments in the laboratory is personally responsible for the setup placed at his disposal and for the experiments performed with it. For experiments during working hours one should try, in the interest of personal safety, to make sure that a second person is present in the testing room. If this is not possible, then at least the times of the beginning and end of an experiment should be communicated to a second person.

When working with high-voltages outside working hours, a second person familiar with the experimental setups must be present in the same room.

If several persons are working with the same setup, they must all know who is to perform the switching operations for a particular experiment. Before switching high-voltage setups on, warning should be given either by short horn signals or by the call: "Attention! Switch-on!" This is especially important during loud experiments, so that people standing by may cover their ears. If necessary, switching off can be announced after completion either by a single long hooting tone or by the call: "Switched off".

Explosion and Fire Risk, Radiation Protection

In experiments with oil and other easily inflammable materials, special care is necessary owing to the danger of explosion and fire. In each room where work is carried out with these materials, suitable fire extinguishers must be to hand, ready for use. Easily inflammable waste products, e.g. paper or used cotton waste, should always be disposed of immediately in metal bins. Special regulations must be observed when radioactive sources are used.

Accident Insurance

Everyone working in the Institute must be insured against accidents.

Conduct During Accidents

Mode of action in the case of an electrical accident:

1. Switch off the setup on all poles. So long as this has not been done, the victim of the accident should not be touched under any circumstances.
2. If the victim is unconscious, notify the life-saving service at once: Telephone ... : Immediate attempts to restore respiration by artificial respiration, mouth-to-mouth respiration or chest massage! These measures must be continued, if necessary, up to the beginning of an operation. (Only 6 to 8 minutes time before direct heart massage!)
3. Even during accidents with no unconsciousness, it is recommended that the victim lie quietly and a doctor's advice be sought.

Appendix 2

Calculation of the Short-Circuit Impedance of Transformers in Cascade Connection

In the general case the stages of a cascade according to Fig. 1.1-2 consist of three windings, the potentials of which are independent of one another. This condition is fulfilled by the equivalent circuit in Fig. A 2-1, where an impedance $\widetilde{Z}_E$, $\widetilde{Z}_H$ or $\widetilde{Z}_K$ is attributed to each winding; each impedance is connected in series with an ideal 3-winding transformer with the corresponding number of turns N_E, N_H or N_K. The impedances are determined either from calculated or experimentally derived results of the three short-circuit tests between any two windings taken at a time [e.g. *Siemens* 1960].

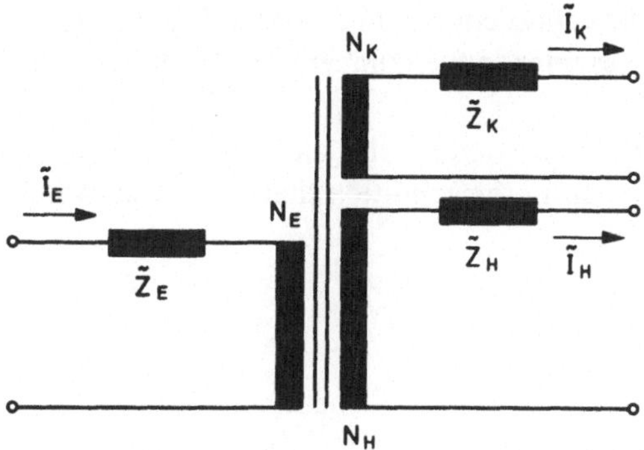

Fig. A 2-1
Equivalent circuit of one stage of a cascade

For each stage it follows that when the magnetizing current is neglected, the sum of the ampere-turns of all the windings must be equal to zero:

$$N_E \tilde{I}_E - N_H \tilde{I}_H - N_K \tilde{I}_K = 0.$$

The method of calculation shall be illustrated by the example of a 3-stage cascade, where the losses shall be neglected for the sake of clarity:

$$\tilde{Z}_E = j X_E; \quad \tilde{Z}_H = j X_H; \quad \tilde{Z}_K = j X_K.$$

Further it will be assumed that the ratio of the number of turns is the same for all stages, viz:

$$N_E / N_H = N_K / N_H.$$

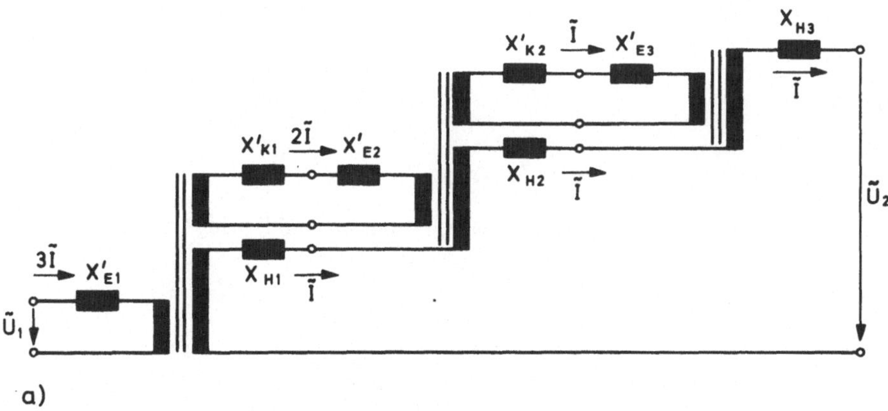

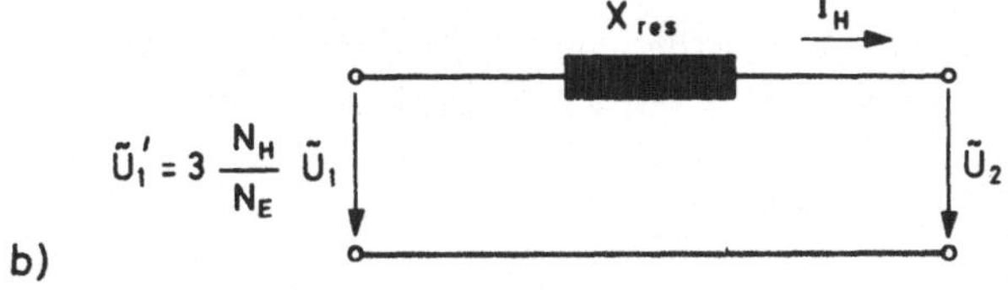

Fig. A 2-2
Equivalent circuits of a 3-stage cascade
a) Complete equivalent circuit
b) Simplified equivalent circuit

The result of the assumptions made above is the equivalent circuit shown in Fig. A 2-2a. The indicated currents and the dashed reactances refer to the number of turns N_H of the respective high-voltage winding.
An equivalent circuit as in Fig. A 2-2b shall now be derived for the entire cascade. The resulting short-circuit reactance X_{res} is obtained from the condition that the power rating be the same:

$$I_H^2 X_{res} = \sum_{\nu=1}^{3} (I'^2_{E\nu} X'_{E\nu} + I'^2_{K\nu} X'_{K\nu} + I^2_{H\nu} X_{H\nu}).$$

From this it follows at once that:

$$X_{res} = X_{H1} + X_{H2} + X_{H3} + X'_{K2} + X'_{E3} + 4(X'_{K1} + X'_{E2}) + 9X'_{E1}.$$

With the simplifications as before, one obtains for the short-circuit reactance of an n-stage cascade:

$$X_{res} = \sum_{\nu=1}^{n} [X_{H\nu} + \nu^2(X'_{E(n+1-\nu)} + X'_{K(n-\nu)})]$$

with $X'_{K_0} = 0$.
This method is not bound to the simplifications made here for the sake of clarity; it can easily be extended to different transformer ratios and can also take the effective resistances into account. It may also be used for the calculation of the short-circuit impedance of potential transformers in cascade connection.

Appendix 3

Calculation of Single-Stage Impulse Voltage Circuits

For the circuit b as in Fig. 1.3-3b the following equations are valid, using the same notations:

$$U_0 - \frac{1}{C_s}\int_0^t (i_e + i_d)\,dt = i_e R_e = i_d R_d + u(t)$$

$$i_d = C_b \frac{du(t)}{dt} \quad \text{with } u(t=0) = 0.$$

This differential equation will be solved by applying the Laplace transformation. For the functions in the p-plane, the corresponding capital letters will be used as symbols:

$$\frac{U_0}{p} - \frac{1}{pC_s}[J_e + pC_b U] = J_e R_e = U(pR_d C_b + 1).$$

Solving for $U = U(p)$, we have:

$$U(p) = \frac{U_0}{R_d C_b} \frac{1}{p^2 + bp + c}$$

where

$$b = \frac{1}{R_e C_s} + \frac{1}{R_d C_s} + \frac{1}{R_d C_b}$$

$$c = \frac{1}{R_d C_b R_e C_s} \; .$$

The two roots of the quadratic equation in the denominator polynomial are:

$$p_{1,2} = \frac{b}{2} \left(- 1 \pm \sqrt{1 - \frac{4c}{b^2}} \right).$$

These are always < 0 and real. Reverse transformation into the t-plane gives:

$$u(t) = \frac{U_0}{R_d C_b} \frac{1}{p_1 - p_2} (e^{p_1 t} - e^{p_2 t})$$

$$u(t) = \frac{U_0}{R_d C_b} \frac{T_1 T_2}{T_1 - T_2} (e^{-t/T_1} - e^{t/T_2}).$$

Here the time constants $T_1 = - 1/p_1$ and $T_2 = - 1/p_2$ have been introduced. The general solution can be appreciably simplified if the usually valid approximation

$$R_e C_s \gg R_d C_b$$

is considered. Then the relationships

$$b \approx \frac{1}{R_d} \left(\frac{C_s + C_b}{C_s C_b} \right) \quad \text{and} \quad \frac{4c}{b^2} \ll 1$$

follow. With that the square root expression in $p_{1,2}$ approaches the value $(1 - 2c/b^2)$, and it follows that

$$p_1 \approx - \frac{c}{b} = - \frac{1}{R_e (C_s + C_b)} , \qquad T_1 = R_e (C_s + C_b)$$

and

$$p_2 \approx \frac{c}{b} - b \approx - b = - \frac{1}{R_d} \left(\frac{C_s + C_b}{C_s C_b} \right) , \qquad T_2 = R_d \left(\frac{C_s C_b}{C_s + C_b} \right).$$

Appendix 4

Calculation of the Impedance of Plane Conductors

In high-voltage setups rapidly varying high currents are often passed through extended plane conductors. In choosing the dimensions of these conductor systems the problem arises of determining the voltage drops which occur or the impedances responsible for them. To do this however, it is first necessary to reach an agreement about what shall be considered an impedance. This shall be done on the basis of Fig. A 4-1.

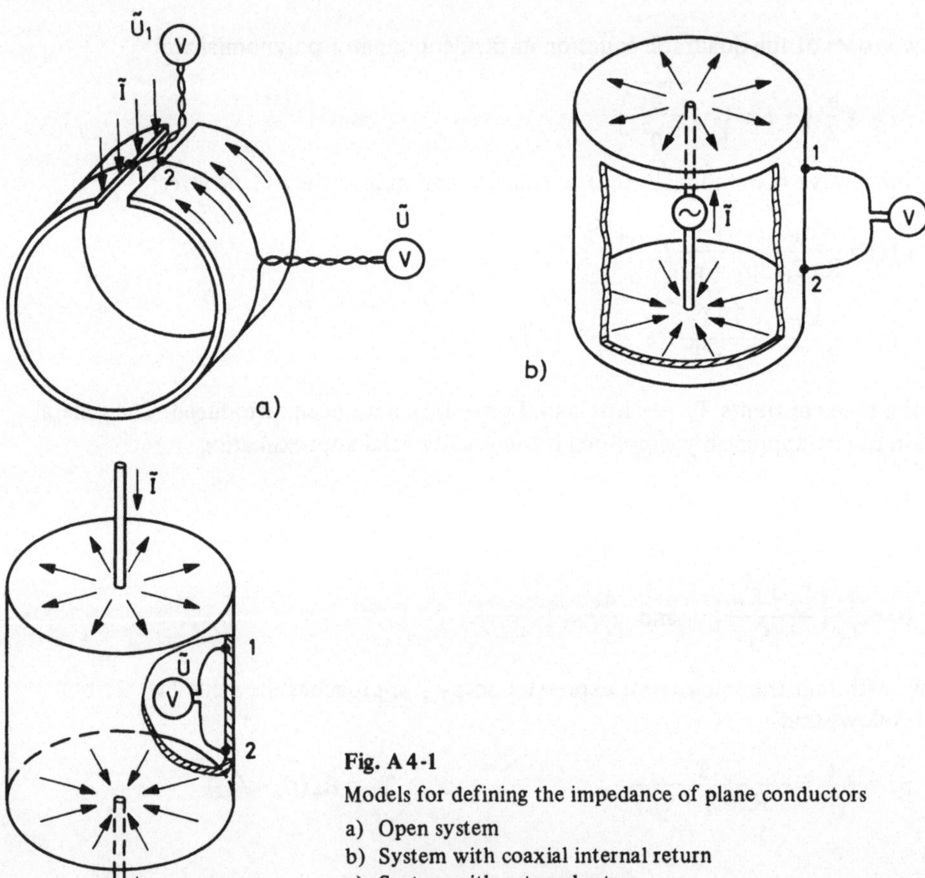

Fig. A 4-1
Models for defining the impedance of plane conductors
a) Open system
b) System with coaxial internal return
c) System with external return

As an example of an open system as in a) consider a conductor band formed into a circular cylinder carrying an alternating current $\tilde{I}$, as shown by the indicated arrows. On account of the enclosed magnetic flux a voltage $\tilde{U}_1$ would be measurable across the shortest distance between points 1 and 2 even if the band were an ideal conductor. Decisive for the impedance of the conductor band can therefore sensibly only be that voltage $\tilde{U}$ which would be measured by a circumferential arrangement of the voltage measuring leads:

$$\tilde{Z} = \frac{\tilde{U}}{\tilde{I}} .$$

Moreover, for practical cases $\tilde{U}$ can also be considered as that voltage by which the voltage $\tilde{U}_1$ is greater than in the case of an ideal conductor. In closed systems, such as the cylindrical chamber in Fig. A 4-1b, no magnetic field occurs on the outside, which is why the arrangement of the voltage measuring leads to determine the voltage $\tilde{U}$ between points 1 and 2 can be arbitrary. Nevertheless, it should be observed here that for high angular frequencies $\omega = 2\pi f$ the current density on the external surface of plate-type plane conductors, consisting of metallic foils for example, will be very low, and then no potential difference would be measurable from the outside. This situation occurs when the depth of penetration

$$\delta = \frac{1}{\sqrt{\pi \mu \kappa f}}$$

is small compared with the thickness w of the plate-type plane conductor. Here $\mu = \mu_0 \mu_r$ denotes the permeability and κ the specific conductivity. The following are some guiding values for copper and iron ($\mu_r = 200$):

f = 1 MHz: $\delta_{Cu} = 0.07$ mm f = 100 kHz: $\delta_{Cu} = 0.21$ mm
$\delta_{Fe} = 0.01$ mm $\delta_{Fe} = 0.04$ mm

An application of the system as shown in Fig. A 4-1b is the measurement of the voltage drop in the walls of a completely screened laboratory. Fig. A 4-1c shows a system which also has a hollow cylindrical conductor, as used in foil measuring resistors.

From a current-carrying plane conductor a small square can always be cut such that the current paths on two opposite sides are parallel. Its impedance is defined as the specific plane impedance:

$$\tilde{Z}' = R' + j\omega L'.$$

It is identical to the impedance of a square plane conductor of arbitrary length of side carrying parallel component currents, as would be the case for $l = b$ in Fig. A 4-2. This is true for the plate-type plane conductor shown under a) as well as for the lattice-type shown under b). In the latter case the measured result is independent of the angle β between the lattice bars and the sides of the square, so long as the spacing between the component conductors is $a \ll b$.

For the specific plane impedances of the most important types of conductor, the relationships summarized below are valid for plate-type plane conductors of thickness w [*Lautz* 1969]:

for $w \ll \delta$ it follows that $R' = \frac{1}{\kappa w}$; $L' = \frac{\mu w}{8}$,

for $w \gg \delta$, assuming double-sided current flow, it follows that

$$R' = \frac{1}{\kappa\, 2\delta}; \quad L' = \frac{\mu\, 2\delta}{8} = \frac{R'}{\omega} .$$

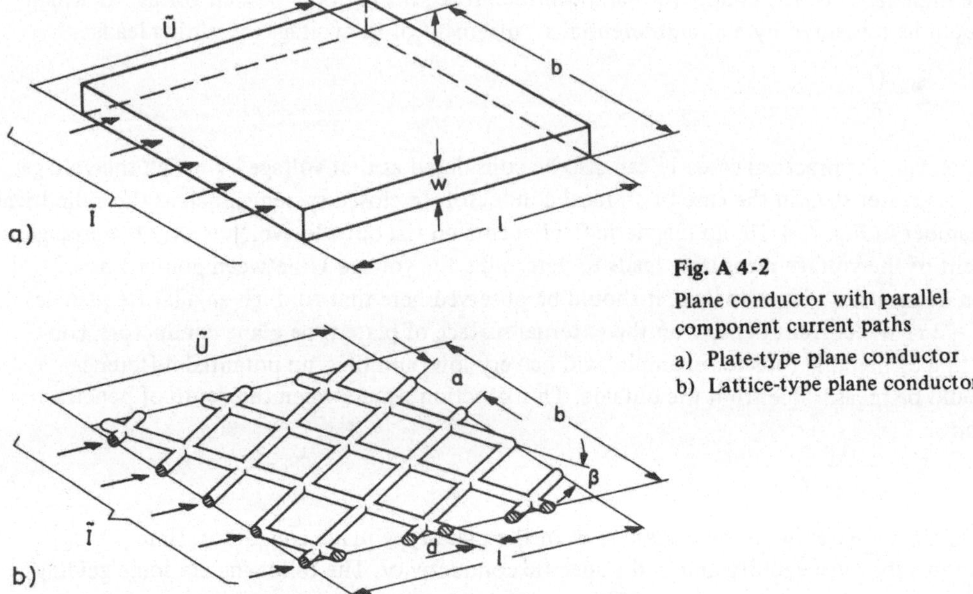

Fig. A 4-2
Plane conductor with parallel component current paths
a) Plate-type plane conductor
b) Lattice-type plane conductor

For single-sided current flow the effective thickness of the conducting layer is halved and therefore

$$R' = \frac{1}{\kappa\,\delta}\,; \quad L' = \frac{\mu\,\delta}{2} = \frac{R'}{\omega}\,.$$

Lattice-type plane conductors with wire diameter d and mean spacing of wires $a \gg d$ [*Sirait* 1967; *Hylten-Cavallius, Giao* 1969]:

for $d \ll \delta$ it follows

$$R' = \frac{4a}{\kappa\,\pi\,d^2}\,; \quad L' = L_i' + L_a'$$

$$L_i' = \frac{\mu\,a}{8\pi}$$

$$L_a' = -\frac{\mu_0\,a}{2\pi}\ln\left(\sin\frac{\pi\,d}{2a}\right),$$

for $d \gg \delta$ it follows

$$R' = \frac{a}{\kappa\,\pi\,d\delta}\,; \quad L' = L_i' + L_a'$$

$$L_i' = \frac{1}{\omega}\,\frac{a}{\kappa\,\pi\,d\delta} = \frac{R'}{\omega}$$

L_a' as for $d \ll \delta$.

Under the condition that a ≪ d these relationships are also valid when instead of the lattice-type plane conductor with the square lattice shown in Fig. A 4-2b, other forms are considered, such as hexagonal lattices. For plane conductors with parallel component currents, the impedance of a suitably chosen rectangle of breadth b and length l as in Fig. A 4-2 is given by:

$$\tilde{Z} = \tilde{Z}' \frac{l}{b} .$$

For plane conductors with radial component currents the impedance of a suitably chosen circular ring with radii r and R as in Fig. A 4-3 can be calculated as:

$$\tilde{Z} = \tilde{Z}' \frac{1}{2\pi} \ln \frac{R}{r}$$

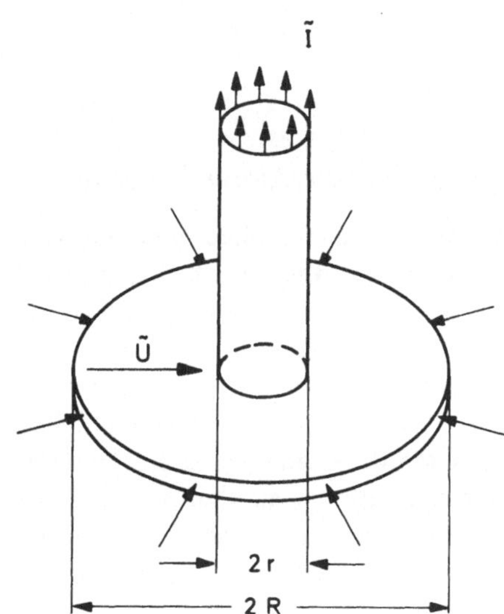

Fig. A 4-3
Plane conductor with a single current junction

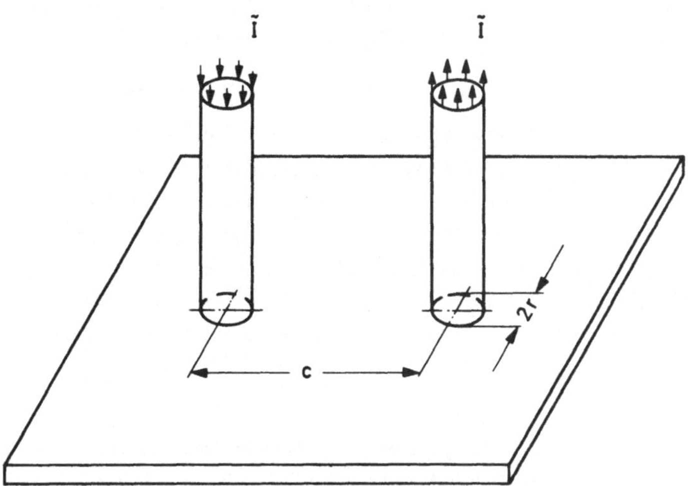

Fig. A 4-4
Plane conductor with double current junction

It is often required to know the potential difference between two current junctions in an extended plane conductor. Such an arrangement is shown in Fig. A 4-4, where it is assumed that the currents are led into the plane conductor by cylindrical conductors of radius r at a mean spacing c; the dimensions of the plane conductor are very large compared with c. Analogous to the capacitance of a double line [e.g. *Lautz* 1969], we have

$$\tilde{Z} = \tilde{Z}' \frac{1}{\pi} \ln(x + \sqrt{x^2 - 1}), \quad \text{with} \quad x = c/2r.$$

Appendix 5

Statistical Evaluation of Measured Results

For the experimental determination of the electric strength of insulation systems one obtains measured values which can show appreciable dispersion, depending upon the insulating material and the electrode configuration. If the fluctuations of the measured values are of a random nature, then it is appropriate to apply the methods of mathematical statistics to their evaluation[1]). In this way it is possible to make statements of reliable certainty about the performance of a large assembly with the aid of only a few measurements. Moreover, the results can be represented in a simple and clear manner and easily compared with one another. In certain cases, using statistical methods, it is possible to show that different mechanisms are at play, for instance when a series of measurements may be divided into various subgroups.

The application of statistics shall be discussed here for the example of the breakdown discharge voltage U_d, which is especially important in high-voltage technology; the same principles apply to other measured quantities. It is useful to distinguish between two groups of results; these shall be treated below under A 5.1 and A 5.2.

A 5.1 Direct Determination of Probability Values (Series Stressing)

In a first group of investigations a voltage of given time dependence is repeatedly applied to the same sample (or to several identical samples for destructive breakdown discharges); the number of breakdown discharges n_d out of a total number of applications n is determined each time for a specified value of the voltage U. The breakdown discharge probability $P(U) = n_d/n$ is thus directly obtained. For example, when testing insulators with full impulse voltages, the distribution function of the breakdown discharge voltage shown in Fig. A 5-1 is directly obtained. Some important characteristic values are the voltages U_{d-50}, U_{d-5} and U_{d-95}.

[1]) This appendix was compiled on the basis of the IEC-Publication 60-2 (1973): High-Voltage Test Techniques, Test Procedures; further literature, e.g. *Kreyszig* 1967; DIN 1319; *Rasquin* 1972.

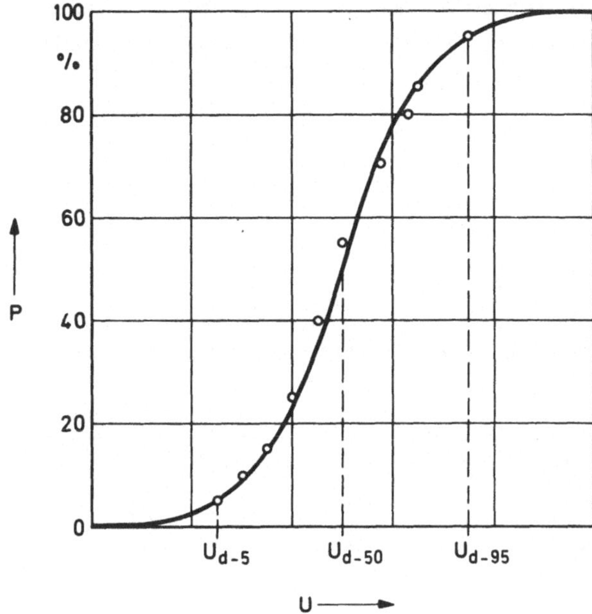

Fig. A 5-1
Experimental distribution function for breakdown discharge voltages, plotted in a linear coordinate system

For the evaluation the measured values of breakdown discharge probabilities for different voltages are suitably plotted in a probability net and one obtains the result shown in Fig. A 5-2. If the measured points lie approximately on a straight line as indicated in the figure, it may be assumed that the breakdown discharge voltage of the investigated sample obeys a Gaussian or Normal distribution law. That is to say the ordinate of this probability net is so divided that the cumulative frequency curve of a Normal distribution becomes a straight line. The assumption of a Normal distribution for the breakdown discharge voltage of electrode arrangements with gaseous, liquid or solid insulation is permissible in most cases, provided one restricts oneself to the range of about 5 ... 95 % breakdown discharge probabilities; outside this range special methods must be adopted [see, for example, IEC Publication 60-2 (1973); High-Voltage Test Techniques, Test Procedures].

Once the straight line approximating the measured points has been drawn in the probability net, the value $\bar{U}_d \approx U_{d-50}$ is read off at the breakdown discharge probability P(U) = 50 %. Further, the standard deviation s of the measured series is obtained as the difference of the voltages at P(U) = 50 % and 16 %, or also 50 % and 84 %, since the Gaussian distribution is symmetrical.

A 5.2 Determination of the Distribution Function of a Measured Quantity

In a second group of investigations a certain voltage is applied to a sample and increased until a breakdown or flashover occurs. In a subsequent experiment on the same sample (or on another identical sample for destructive breakdown discharges) a slightly different value of the breakdown discharge voltage results. Thus one obtains a series of measured $\bar{U}_d$ values which show some dispersion. Recording the impulse voltage-time characteristics of gaps or surge diverters belongs to this type of test for example. For a series of n breakdown discharge voltage values U_{d_i}, the mean value $\bar{U}_d$ and the standard deviation s are calculated using the following equations:

$$\bar{U}_d = \frac{1}{n} \sum_{i=1}^{n} U_{d_i}$$

$$s = \sqrt{\frac{1}{n-1} \sum_{i=1}^{n} (U_{d_i} - \bar{U}_d)^2}\,.$$

The standard deviation can also be referred to U_d and is then termed the coefficient of variation v:

$$v = \frac{s}{\bar{U}_d}\,.$$

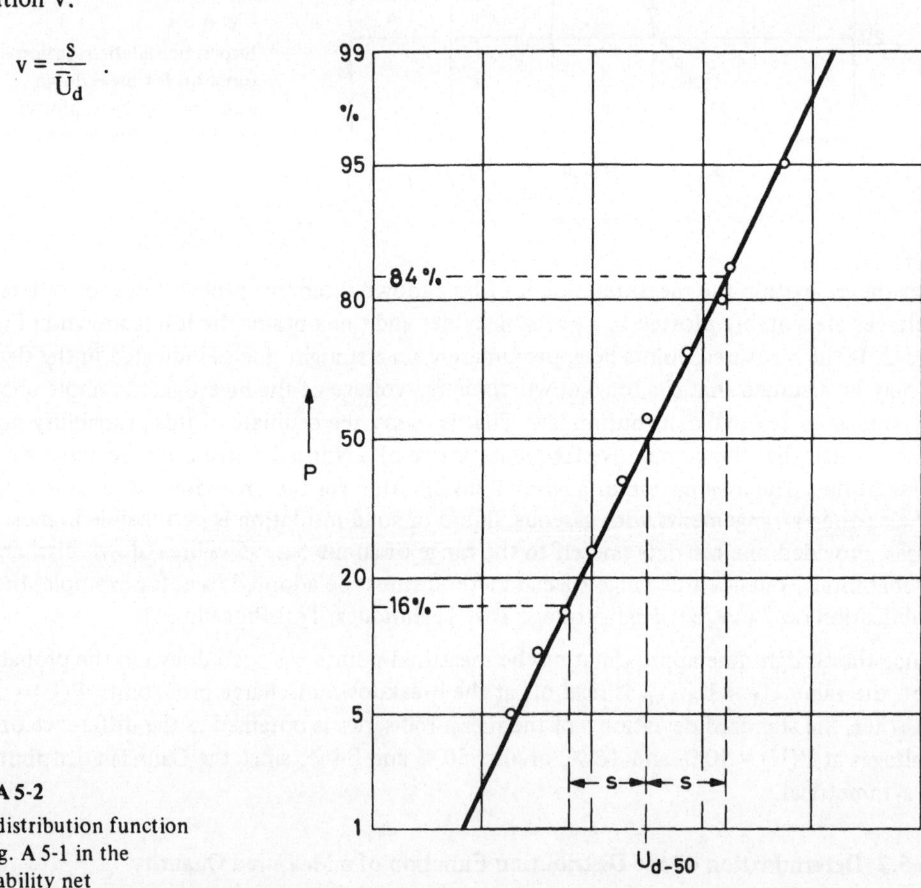

Fig. A 5-2
The distribution function of Fig. A 5-1 in the probability net

This calculation can be performed quite generally for any arbitrary distribution. When a Normal distribution is assumed 84 % − 16 % = 68 % of all the U_d values must lie within the range $\bar{U}_d - s$ to $\bar{U}_d + s$.

Alternatively, graphical evaluation of the series on probability paper is also possible, analogous to Fig. A 5-2. Here $P(U_d) = n_i/(n + 1)$ is plotted in the probability net as a function of U_{d_i}, where the breakdown discharge values are arranged according to their magnitude; n_i is the number of disruptive discharges up to and including the voltage U_{d_i}, out of a total of n voltage applications. The distribution function is again approximated in the probability net by a straight line, from which $\bar{U}_d$ and s can be determined as described above. Exactly the same values as calculated from the equations are only obtained in exceptional cases; nevertheless, the graphical method provides a picture of the distribution function which can be extremely informative in many cases.

A 5.3 Determination of the Confidence Limits of the Mean Value of the Breakdown Discharge Voltage

The values $\bar{U}_d$ and s, obtained as in A 5.2 from the limited number of n measurements of a series, in the mathematical sense represent more or less accurate estimates of the corresponding values of the very large total population of samples. Once again, under the assumption of a Normal distribution of U_d values, one can specify the limits for a measurement within which the mean value of a series with $n \to \infty$ can be expected to lie for a given statistical certainty P. The calculation of these "mean value confidence limits" is very useful, particularly for the comparison of different series of measurements.

Consider a random sample of n measured values, the mean value and standard deviation of which were calculated as $\bar{U}_d$ and s respectively. The mean value of the breakdown discharge voltage, determined from an infinitely large number of individual measurements, would then lie for a given certainty P, within the confidence limits of the mean value of the random sample comprising n test samples, viz:

$$\bar{U}_d \pm \frac{t}{\sqrt{n}}\, s\,.$$

The factor t depends upon the value chosen for P as well as on the random sample number n and is tabulated in statistical handbooks [*Owen* 1962; *Kreyszig* 1967]. For a statistical certainty of P = 95 %, the following values may be quoted:

n	5	10	20	50	100	200	∞
t	2.8	2.3	2.1	2.0	2.0	1.97	1.96
$t/\sqrt{n}$	1.24	0.72	0.47	0.28	0.2	0.14	0

Should one wish, on the basis of random sampling for example, to determine which of two slightly different types of test sample has the higher electrical strength, then the mean value $\bar{U}_d$ and its confidence limits are calculated for each random sample. If the confidence intervals of both samples do not overlap, one may then assert that, for the chosen statistical certainty of, for instance, P = 95 %, the one model has a higher breakdown discharge voltage. This is represented graphically in Fig. A 5-3. If the confidence intervals overlap by more than a quarter of the smaller interval, then the measured difference could be incidental [*Sachs* 1970]. An example of this is shown in Fig. A 5-4.

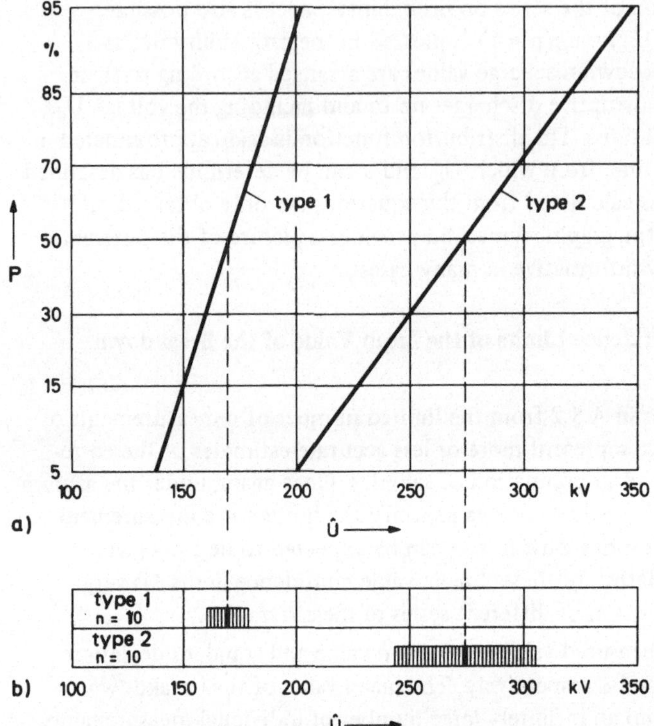

Fig. A 5-3
Impulse breakdown discharge voltages of two different types of test object

a) distribution functions
b) confidence limits of mean values U_{d-50} (P = 95 %); an effect of the design on U_{d-50} is statistically ensured

A 5.4 Details for the Determination of Breakdown Discharge Voltages with a Given Probability

If the mean value and standard deviation of breakdown discharge voltages of an electrode configuration are known, statements about the probable distribution of the measured values can be made, once again assuming a Normal distribution. Out of n = 1000 independent individual measurements,

317 lie outside the range $\bar{U}_d \pm s$,
46 lie outside the range $\bar{U}_d \pm 2s$,
3 lie outside the range $\bar{U}_d \pm 3s$,

where it is assumed that for n = 1000 the measured values of $\bar{U}_d$ and s differ only minimally from the corresponding values for the total population. To be precise, and particularly for a small random sample number, the confidence limits of $\bar{U}_d$ and s must be taken into consideration.

In practical measurements, as in A 5.1, the value $\bar{U}_d - 3s$ is often used as the estimated value for the impulse withstand voltage U_{d-0} of an electrode configuration; $\bar{U}_d + 3s$ is then the assured breakdown discharge voltage U_{d-100}. In the case of a Normal distribution of the breakdown discharge voltages, and with sufficiently accurate values of $\bar{U}_d$ and s, these limiting values correspond to a breakdown discharge probability of 0.14 or 99.86 % respectively.

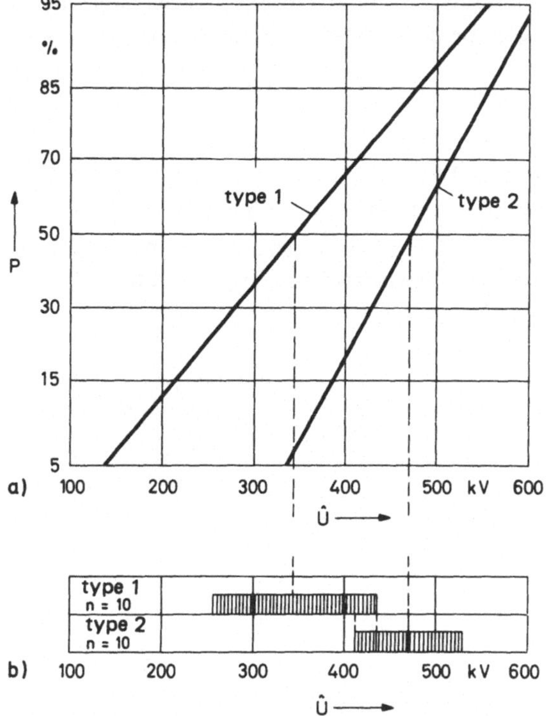

Fig. A 5-4 Impulse breakdown discharge voltages of two different types of test object

a) distribution functions

b) confidence limits of mean values U_{d-50} (P = 95 %); no statistically ensured effect of the design on U_{d-50}

A 5.5 "Up and Down" Method for Determining the 50 % Breakdown Discharge Voltage

When only the 50 % breakdown discharge voltage of an electrode configuration, as discussed under A 5.1, is to be determined with a minimum sacrifice of time and yet to good accuracy, the "up and down" method is especially well suited for this purpose. With only a small number of measurements this method supplies a very good estimate of U_{d-50}.

Initially one chooses a voltage U_k (an estimated value of the required breakdown discharge voltage), and a voltage interval ΔU_k which should be about 3 % of U_k. An impulse voltage with a peak value U_k is then applied to the sample. If no breakdown or flashover occurs the next impulse is given the peak value $U_k + \Delta U_k$. If a breakdown discharge occurs the next peak value is $U_k - \Delta U_k$. This process is continued, whereby the peak value of each impulse voltage is determined by the result of the preceding experiment. The number of impulse voltages n_i for every peak value U_i obtained as in the above procedure is recorded; the 50 % breakdown discharge voltage can then be determined using the following equation:

$$U_{d-50} = \frac{\Sigma n_i U_i}{\Sigma n_i} .$$

Indeed, even for $\Sigma n_i = 20$ the value so determined lies, to a high degree of certainty, within the range of breakdown discharge probability between P(U) = 30 % and P(U) = 70 %. The standard deviation can also be obtained from this kind of series for determining U_{d-50} [*Dixon, Massey* 1969]. However, a larger number of measurements Σn_i would then be necessary.

References

AEG, Das Hochspannungsinstitut der AEG, Festschrift zur Eröffnung des Instituts in Kassel 1953, s. auch AEG-Mitteilungen **43** (1953), S. 256–304.

Anderson, J. C., Dielectrics, Chapman and Hall, London 1964.

Anis, H., Trinh, N. G., Train, D., Generation of Switching Impulses using High-Voltage Testing Transformers. IEEE Trans. Power Apparatus and Systems, Vol. PAS-94 (1975), Pg. 187–195.

Baatz, H., Überspannungen in Energieversorgungsnetzen, Springer, Berlin 1956.

Baldinger, E., Kaskadengeneratoren, in Flügge, S.: Handbuch der Physik, Bd. 44, Springer Berlin 1959.

Bertele, H., Mitterauer, J., Hochstromtechnik in der modernen Forschung und Entwicklung der Kernfusion, E. u. M. **87** (1970), S. 139–152 und S. 305–353.

Bewley, L. V., Traveling Waves on Transmission Systems, 2. Aufl., Dover-Publ., New York 1951.

Boeck, W., Eine Scheitelspannungs-Meßeinrichtung erhöhter Meßgenauigkeit mit digitaler Anzeige, ETZ-A **84** (1963), H. 26, S. 883–886.

Böning, P., Das Messen hoher elektrischer Spannungen, Braun, Karlsruhe 1953.

Böning, P., Kleines Lehrbuch der elektrischen Festigkeit, Braun, Karlsruhe 1955.

Carrara, G., Zaffanella, L., UHV Laboratory Clearance Tests, IEEE-Paper Nr. 68 P 692-PWR, Chicago 1968.

Craggs, J. D., Meek, J. M., High Voltage Laboratory Technique, Butterworth, London 1954.

Deutsch, F., Schalter für Hochstromimpulse bei hohen Spannungen, Bull. SEV **55** (1964), Nr. 22, S. 1123–1129.

Dixon, W. J., Massey, F. J., Introduction to Statistical Analysis, McGraw Hill, New York 1969.

Elsner, R., Das neue Höchstspannungsprüffeld der Siemens-Schuckertwerke in Nürnberg, Siemens-Z. **26** (1952), H. 6, S. 259–267.

Felici, N. J., Elektrostatische Hochspannungs-Generatoren, Braun, Karlsruhe 1957.

Fischer, A., Hochspannungslaboratorien im In- und Ausland, ETZ-A **90** (1969), H. 25, S. 656–662.

Fitch, R. A., Howell, V. T. S., Novel Principle of Transient High-Voltage Generation, Proc. IEE **111** (1964), Nr. 4, S. 849–855.

Flegler, E., Einführung in die Hochspannungstechnik, Braun, Karlsruhe 1964.

Früngel, F., Impulstechnik, Akad. Verlagsges., Leipzig 1960.

Früngel, F., High Speed Pulse Technology, Vol. I u. II, Academic Press, New York 1965.

Gänger, B., Der elektrische Durchschlag von Gasen, Springer, Berlin 1953.

Gontscharenko, G. M., Dmochowskaja, L. F., Shakov, E. M., Ispitalelnie Ustanowki; i ismeritelnie Ustroistwa W Laboratorijach wisokogo (Versuchsanlagen und Meßeinrichtungen in Hochspannungslaboratorien), Naprijaschenija, Moskwa 1966.

Grabner, K., 1400-kV-Wechselspannungs-Prüfkaskade in Säulenbauweise, ELIN-Z. 19 (1967), S. 24–32.

Greenwood, A., Electrical Transients in Power Systems, Wiley, New York 1971.

Gsodam, H., Stockreiter, H., Das neu erbaute Hochspannungslaboratorium und Transformatorenprüffeld der ELIN-UNION im Werk Weiz, ELIN-Z. **17** (1965), S. 132–137.

Hartig, A., Unvollkommener und vollkommener Durchschlag in Schwefelhexafluorid, Diss. TH Braunschweig 1966 (Beiheft Nr. 3 der ETZ).

Hayashi, Ch., Nonlinear Oscillations in Physical Systems, McGraw Hill, New York 1964.

Hecht, A., Elektrokeramik, Springer, Berlin 1959.

Heilbronner, F., Das Durchzünden mehrstufiger Stoßgeneratoren, ETZ-A **92** (1971), H. 6, S. 372–376.

Heise, W., Tesla-Transformatoren, ETZ-A **85** (1964), H. 1, S. 1–8.

Heller, B., Veverka, A., Surge Phenomena in Electrical Machines, Akademia Prague 1968.

Helmchen, G., Die Entwicklung der Stoßspannungstechnik, ETZ-A 84 (1963), H. 4, S. 107–113.

Herb, R. G., Van de Graaff Generators, in Flügge, S.: Handbuch der Physik, Bd. 44, Springer Berlin 1959.

Heyne, V., Erweiterung des Transformatorenprüffeldes und Neubau eines Hochspannungslaboratoriums, BBC-Nachr. **51** (1969), H. 2, S. 67–73.

v. Hippel, A., Dielectric Materials and Applications, 2. Aufl. Wiley, New York 1958.

Hövelmann, F., Untersuchungen über das Stoßdurchschlagsverhalten von technischen Elektrodenanordnungen in Luft von Atmosphärendruck, Diss. TH Aachen 1966.

Hylten-Cavallius, N. R., Giao, T. N., Floor Net Used as Ground Return in High-Voltage Test Areas, IEEE PAS **88** (1969), Nr. 7, S. 996–1005.

Hylten-Cavallius, N. R., Calibration and Checking Methods of Rapid High-Voltage Impulse Measuring Circuits, IEEE PAS **89** (1970), Nr. 7, S. 1393–1403.

Imhof, A., Hochspannungs-Isolierstoffe, Braun, Karlsruhe 1957.

Jiggins, A. H., Bevan, J. S., Voltage Calibration of a 400 kV Van de Graaff Machine, J. Sc. Instruments **43** (1966), S. 478–479.

Kaden, H., Wirbelströme und Schirmung in der Nachrichtentechnik, 2. Aufl., Springer Berlin 1959.

Kappeler, H., Hartpapierdurchführungen für Höchstspannung, Bull. SEV **40** (1949), Nr. 21, S. 807–815.

Kieback, D., Der Oeldurchschlag bei Wechselspannung, Diss. TU Berlin 1969.

Kind, D., Meßgerät für hohe Spannungen mit umlaufenden Meßelektroden, ETZ-A 77 (1956), H. 1, S. 14–16.

Kind, D., Die Aufbaufläche bei Stoßspannungsbeanspruchung technischer Elektrodenanordnungen in Luft, ETZ-A 79 (1958), H. 3, S. 65–69.

Kind, D., Salge, J., Über die Erzeugung von Schaltspannungen mit Hochspannungs-Prüftransformatoren, ETZ-A **86** (1965), H. 20, S. 648–651.

Kind, D., König, D., Untersuchungen an Epoxidharzprüflingen mit künstlichen Hohlräumen bei Wechselspannungsbeanspruchung, Elektrie **21** (1967), S. 9–13.

Knoepfel, H., Pulsed High Magnetic Fields, North-Holland Publ. Comp., Amsterdam 1970.

Knudsen, N. H., Abnormal Oscillations in Electric Circuits Containing Capacitance, Trans. of the Royal Institute of Technology Nr. 69, Stockholm 1953.

Kodoll, W., Teilentladungs-Durchschlag von polymeren Isolierstoffen bei Wechselspannung, Diss. TU Braunschweig 1974.

Kreuger, F. H., Discharge Detection in High-Voltage Equipment, Heywood, London 1964.

Kreyszig, E., Statistische Methoden und ihre Anwendungen, Vandenhoeck & Ruprecht, Göttingen 1967.

Küpfmüller, K., Einführung in die theoretische Elektrotechnik, Springer Berlin 1965.

Kuffel, E., Abdullah, M., High-Voltage Engineering, Pergamon Press, Oxford 1970.

Läpple, H., Das Versuchsfeld für Hochspannungstechnik des Schaltwerks der Siemens-Schuckert-Werke, Siemens-Z. **40** (1966), H. 5, S. 428–435.

Lautz, G., Elektromagnetische Felder, Teubner, Stuttgart 1969.

Leroy, G., Bouillard, J. G., Gallet, G., Simon, M., Essais diélectriques et très hautes tensions „Le L.T.H.T. des Renardières", Société Française des Electriciens, 1ère Section, Avril 1971.

Leroy, G., Gallet, G., Eléments pour un projet de laboratoire à haute tension, E.D.F. bulletin de la direction des études et recherches – série B, réseaux électriques, matériels électriques Nr. 3/4 (1975), Pg. 5–44.

Lesch, G., Lehrbuch der Hochspannungstechnik, Springer Berlin 1959.

Leschanz, A., Oberdorfer, G., Das neue Hochspannungsinstitut der Technischen Hochschule Graz, Elektrotechnik und Maschinenbau **85** (1968), S. 527–532.

Leu, J., Teilentladungen in Epoxidharz-Formstoff mit künstlichen Fehlstellen, ETZ-A **87** (1966), S. 659–665.

Liebscher, F., Held, H., Kondensatoren, Springer Berlin 1968.

Llewellyn-Jones, F., Ionisation and Breakdown in Gases, Science Paperbacks, London 1957.

Lührmann, H., Fremdfeldbeeinflussung kapazitiver Spannungsteiler, ETZ-A **91** (1970), H. 6, S. 332–335.

Lührmann, H., Rasch veränderliche Vorgänge in räumlich ausgedehnten Hochspannungskreisen, Diss. TU Braunschweig 1973.

Marx, E., Hochspannungspraktikum, 2. Aufl., Springer Berlin 1952.

Matthes, W., Zahorka, R., Verzerrung der Spannungskurvenform von Prüftransformatoren infolge Oberschwingungen im Magnetisierungsstrom, ETZ-A **80** (1959), S. 649–653.

Meek, J. M., Craggs, J. D., Electrical Breakdown of Gases, Clarendon Press, Oxford 1953.

Micafil, Hochspannungs-Laboratorium Micafil, Firmenschrift zur Einweihung des Laboratoriums in Zürich 1963.

Minkner, R., Der Drahtwiderstand als Bauelement für die Hochspannungstechnik und Rechentechnik, Meßtechnik **4** (1969), S. 101–106.

Möller, K., Spannungsabfälle in den Wänden metallisch abgeschirmter Hochspannungslaboratorien, ETZ-A **86** (1965), H. 13, S. 421–426.

Mole, G., Basic Characteristics of Corona Detector Calibrators, IEEE PAS **89** (1970), S. 198–204.

Mosch, W., Die Nachbildung von Schaltüberspannungen in Höchstspannungsnetzen durch Prüfanlagen, Wiss. Z. TU Dresden **18** (1969), H. 2, S. 513–517.

Müller, W., Untersuchungen der Spannungsform von Prüftransformatoren an einem Modell, Siemens Zeitschrift **35** (1961), S. 50–57.

Mürtz, H., Hochspannungs-Explosionsverformung, ETZ-B **16** (1964), H. 18, S. 529–535.

Nasser, E., Heiszler, M., Educational Laboratories in High-Voltage Engineering, IEEE E-**12** (1969), Nr. 1, S. 60–66.

Nasser, E., Fundamentals of Gaseous Ionization and Plasma Electronics, Wiley, New York 1971.

Owen, D. B., Handbook of Statistical Tables, Addison-Wesley, London 1962.

Paasche, P., Hochspannungs-Messungen, VEB Verlag Technik, Berlin 1957.

Petersen, C., Untersuchungen über die Zündverzugszeit von Dreielektroden-Funkenstrecken, ETZ-A **86** (1965), H. 17, S. 545–552.

Pfestorf, G. K. M., Jayaram, B. N., Über die theoretische Behandlung der Kaskadenschaltung von Hochspannungstransformatoren, Jahrbuch der TH-Hannover 1958/60.

Philippow, E., Nichtlineare Elektrotechnik, Akd. Verlagsanstalt, Leipzig 1963.

Philippow, E., Taschenbuch Elektrotechnik, Band 2, Starkstromtechnik, VEB Verlag Technik, Berlin 1966.

Philippow, E., Taschenbuch Elektrotechnik, Band 1, Grundlagen, VEB Verlag Technik, Berlin 1968.

Potthoff, K., Widmann, W., Meßtechnik der hohen Wechselspannungen, Vieweg, Braunschweig 1965.

Prinz, H., Zaengl, W., Ein 100-kV-Experimentierbaukasten, Elektrizitätsw. **59** (1960), H. 20, S. 728–734.

Prinz, H., Hochspannungs-Messung mit dem rotierenden Voltmeter, ATM Blatt J 763-3, 4, 5 (1939).

Prinz, H., Feuer, Blitz und Funke, Bruckmann, München 1965.

Prinz, H., Hochspannungsfelder, Oldenbourg, München 1969.

Raether, H., Electron Avalanches and Breakdown in Gases, Butterworth, London 1964.

Rasquin, W., Statistische Auswertung der Meßergebnisse von Durchschlag-Untersuchungen, Bull. SEV **63** (1972), H. 5, S. 231–239.

Raupach, F., MWB-Hochspannungslaboratorium, Firmenschrift zur Inbetriebnahme des Laboratoriums in Bamberg 1969.

Rieder, W., Plasma und Lichtbogen, Vieweg, Braunschweig 1967.

Rodewald, A., Ausgleichsvorgänge in der Marxschen Vervielfachungsschaltung nach der Zündung der ersten Schaltfunkenstrecke, Bull. SEV **60** (1969), H. 2, S. 37–44.

Roth, A., Hochspannungstechnik, 4. Aufl. Springer, Wien 1959.

Rüdenberg, R., Elektrische Schaltvorgänge, 4. Aufl., Springer, Berlin 1953.

Rüdenberg, R., Elektrische Wanderwellen, 4. Aufl., Springer, Berlin 1962.

Sachs, L., Statistische Methoden — Ein Soforthelfer, Springer, Berlin 1970.

Salge, J., Peier, D., Brilka, R., Schneider, D., Application of Inductive Energy Storage for the Production of Intense Magnetic Fields, Proc. 6th Symp. on Fusion Technology, Aachen 1970.

Salge, J., Drahtexplosionen in induktiven Stromkreisen, Habilitationsschrift TU Braunschweig 1971.

Schiweck, L., Untersuchungen über den Durchschlagsvorgang in Epoxidharz-Formstoff bei hohen Spannungen, ETZ-A **90** (1969), H. 25, S. 675–678.

Schwab, A. J., High-Voltage Measurement Techniques, MIT Press, Cambridge Mass. USA (1972).

Siemens, Formel- und Tabellenbuch für Starkstrom-Ingenieure, 2. Aufl., Girardet, Essen 1960.

Sirait, T., Elektrische Ausgleichsvorgänge in den Erdflächenleitern von Hochspannungslaboratorien, Diss. TH Braunschweig 1967.

Sirotinski, L. I., Hochspannungstechnik, Band I, Teil 1: Gasentladungen, VEB Verlag Technik, Berlin 1955.

Sirotinski, L. I., Hochspannungstechnik, Band I, Teil 2: Hochspannungsmessungen, Hochspannungslaboratorien, VEB Verlag Technik, Berlin 1956.

Sirotinski, L. I., Hochspannungstechnik: Äußere Überspannungen, Wanderwellen, VEB Verlag Technik, Berlin 1965.

Sirotinski, L. I., Hochspannungstechnik: Innere Überspannungen, VEB Verlag Technik, Berlin 1966.

Slamecka, E., Prüfung von Hochspannungs-Leistungsschaltern, Springer, Berlin 1966.

Stamm, H., Porzel, R., Elektronische Meßverfahren, VEB Verlag Technik, Berlin 1969.

Stephanides, H., Grundregeln für den Aufbau von Erdungssystemen in Hochspannungslaboratorien, E. u. M. **76** (1959), S. 73–79.

Strigel, R., Elektrische Stoßfestigkeit, Springer, Berlin 1955.

Thione, L., Kučera, J., Weck, K. H., Switching Impulse Generation Techniques using High-Voltage Testing Transformers, Electra, No. **43** (1975), Pg. 33–72.

Unger, H. G., Theorie der Leitungen, Vieweg, Braunschweig 1967.

Wehinger, H., Ausgleichsvorgänge in Prüftransformatoren bei der Erzeugung von Schaltstoßspannungen, Diss. TU Braunschweig 1977.

Wellauer, M., Einführung in die Hochspannungstechnik, Birkhäuser, Basel 1954.

Widmann, W., Stoßspannungs-Generatoren, ATM Blatt Z 44-6, 7, 8 (1962).

Wiesinger, J., Einfluß der Frontdauer der Stoßspannung auf das Ansprechverhalten von Funkenstrecken, Bull. SEV **57** (1966), Nr. 6, S. 243–246.

Wiesinger, J., Funkenstrecken unter Blitz- und Schaltstoßspannungen, ETZ-A **90** (1969), H. 17, S. 407–411.

Winkelnkemper, H., Die Aufbauzeit der Vorentladungskanäle im homogenen Feld in Luft, ETZ-A **86** (1965), H. 20, S. 657–663.

Whitehead, S., Dielectric Breakdown of Solids, Clarendon Press, Oxford 1951.

Zaengl, W., Völcker, O., Messung des Scheitelwerts hoher Wechselspannungen, ATM Blatt V 3383-4 (1961).

Zaengl, W., Das Messen hoher, rasch veränderlicher Stoßspannungen, Diss. TH München 1964.

Zaengl, W., Der Stoßspannungsteiler mit Zuleitung, Bull. SEV **61** (1970), Nr. 21, S. 1003–101[illegible]

Subject Index